MORE!
Teaching Fractions
and Ratios for Understanding

Third Edition

MORE!
Teaching Fractions and Ratios for Understanding

In-Depth Discussion and Reasoning Activities

Third Edition

Susan J. Lamon

Routledge
Taylor & Francis Group

LONDON AND NEW YORK

First published 2012
by Routledge

2 Park Square, Milton Park, Abingdon, Oxon OX14 4RN
711 Third Avenue, New York, NY 10017, USA

Routledge is an imprint of the Taylor & Francis Group, an informa business

First issued in hardback 2017

Library of Congress Cataloging in Publication Data
Lamon, Susan J., 1949-
More teaching fractions and ratios for understanding: in-depth discussion and reasoning activities/Susan J. Lamon. – 3rd ed.
p.cm.
Rev. ed. of: More: in-depth discussion of the reasoning activities in "Teaching fractions and ratios for understanding" / Susan J. Lamon. 2006.
1. Fractions–Study and teaching (Elementary) 2. Ratio and proportion–Study and teaching (Elementary) I. Lamon, Susan J., 1949- More. II. Title.
QA137.L34 2011
372.7'2–dc23
2011019890

ISBN: 978-0-415-88613-0 (pbk)
ISBN: 978-1-138-44221-4 (hbk)

Typeset in Aldine401BT
by Exeter Premedia Services Private Limited

To Bill

Contents

Preface

This resource book accompanies *Teaching Fractions and Ratios for Understanding: Essential Content Knowledge and Teaching Strategies for Teachers*. It was originally intended as a scaffold for adults who are reasoning their way through the fraction world for the first time. Reasoning after many years of rule-based computation presents a challenge to one's understanding, logical thinking, problem solving, and ability to communicate! *MORE* still serves that purpose. However, those who have used the previous editions will note several changes in organization and in content.

In this edition, I have included more activities and supplemental problems in each chapter, including more real-world applications. Additional student work has been added for your analysis, and templates for key manipulatives are provided. Finally, following each chapter is a collection of praxis problems geared to the content of the chapter.

MORE is not an answer key; good reasoning should always produce correct answers, but the *process* is the goal. Everyone knows the conventional symbolic representations and algorithms for getting the correct answers. The purpose here is to demonstrate and to help you engage in powerful ways of thinking so that you can, in turn, enhance the mathematical education of your students.

Solutions are offered with this caveat: no solution should be taken as *the* way to think about a situation. *MORE* offers some suggestions, but no effort is made to exhaust all of the possibilities.

## CHAPTER 1

Proportional Reasoning: An Overview

GETTING STARTED

1. If 6 men can build a house in 3 days, then to shorten the time, you will need more men on the job. As the number of men goes up, the number of days goes down. Six men build $\frac{1}{3}$ of the house in one day, so 12 men could build $\frac{2}{3}$ of the house in one day, and 18 men could build $\frac{3}{3}$ of one whole house in a day (assuming that all of the men do an honest day's work). This means that it would take 3 times as many men to complete the job in $\frac{1}{3}$ the time.

2. If 6 chocolates cost \$0.93, then 12 would cost \$1.86 and 24 would cost \$3.72. If 6 cost \$0.93, then 2 cost \$0.31. The cost of 22 candies is \$0.31 less than \$3.72 or \$3.41.

3. John has 3 times as many marbles as Mark, so you can think of the whole set of marbles as 4 equal groups, with 3 of the groups in front of John and 1 group in front of Mark. If there are 32 marbles, each group contains 8 marbles. John has 24 marbles and Mark has 8 marbles.

4. If Mac does twice as much as his brother, he will do $\frac{2}{3}$ of the lawn, while his brother does $\frac{1}{3}$. If it takes Mac 45 minutes to do $\frac{3}{3}$ of the job, then it takes 15 minutes to do $\frac{1}{3}$ and 30 minutes to do $\frac{2}{3}$. Meanwhile, during that 30 minutes, his little brother does the other $\frac{1}{3}$ of the lawn.

5. The more people you have working, the faster the job will get done (assuming, of course, that the boys do not goof off on the job). If 6 boys were given 20 minutes to clean up, then 1 boy should be given 6 times as much time or 120 minutes. Then 9 would each require $\frac{1}{9}$ of the time needed by 1 person: $\frac{120}{9}$ or $\frac{40}{3}$ or $13\frac{1}{3}$ minutes. Another way. If 6 boys can do the job in 20 minutes, then 3 would take

40 minutes. If there are 9 people working, then each set of 3 people does $\frac{1}{3}$ of the job in 40 minutes, so the total time needed is $\frac{40}{3}$ minutes.

6. No answer. Knowing the weight of one player is not helpful in determining the weight of 11 players. People's weights are not related to each other.

7. Every time they put away $7, Sandra pays $2 and her mom pays $5. To get $210, they will need to make their respective contributions 30 times. In all, Sandra will contribute $60 and her mom will contribute $150.

8. If you decrease the number of people doing a job, it will take longer to finish the job. If you have $\frac{1}{3}$ the number of people working, they can get only $\frac{1}{3}$ as much done in the 96 minutes. Each $\frac{1}{3}$ of the job will take them 96 minutes, so they can do $\frac{2}{3}$ of work in 192 minutes and $\frac{3}{3}$ of the job (the whole job) in 288 minutes (4 hours and 48 minutes).

9. Reason down then reason up. The bike can run for 5 minutes on $0.65 worth of fuel, and for 1 minute on $0.13 worth of fuel. It can run for 6 minutes on $0.78 worth of fuel and for 7 minutes on $0.91 worth of fuel.

10. 15 to 1 = 150 to 10. Therefore, decreasing the number of faculty by 8 will give the required ratio.

11. Your shadow (8′) is 1.6 times as tall as you are (5′), so the shadow cast by the telephone pole must be 1.6 times as tall as the pole. If the shadow is 48′ and that is 1.6 times the real height of the telephone pole, the pole must be 30′ tall. Also, the telephone pole's shadow is 6 times as long as yours, so it must be 6 times as tall as you are.

12. In a square, two adjacent sides have the same length. That means that the ratio of the measure of one side to the measure of the other side would be 1. The rectangle that is most square is the one whose ratio of width to length is closest to 1. For the 35″ × 39″ rectangle, the ratio is $\frac{35}{39}$ or about 0.90; for the 22″ × 25″ rectangle, the ratio is $\frac{22}{25}$ or 0.88. This means that the 35″ × 39″ rectangle is most square.

13. Gear A has 1.5 times the number of teeth on gear B. So every time A turns once, B turns 1.5 times. If A makes 5.5 revolutions, B makes 5.5(1.5) or 8.25 revolutions.

14. The density or crowdedness of a town with cars is given by comparing the number of cars to the number of square miles in the town. For town A, the crowdedness is $\frac{12555}{15} = 837\frac{\text{cars}}{\text{sq mi}}$. For town B, it is $\frac{2502}{3} = 834\frac{\text{cars}}{\text{sq mi}}$. For town C, it is $\frac{14212}{17} = 836\frac{\text{cars}}{\text{sq mi}}$. The town least crowded with cars is town B. B must be Birmingham.

15. Several different approaches are given:

   a. In pitcher A, 4 out of 7 total cubes are cranberry; in B, 3 out of 5 total cubes are cranberry. Because $\frac{3}{5} > \frac{4}{7}$, B has a stronger cranberry taste.

   b. In pitcher A, the ratio of cranberry to apple is 4 to 3; in B, the ratio of cranberry to apple is 3 to 2. Because $\frac{3}{2} > \frac{4}{3}$, B has a stronger cranberry taste.

   c. Using common denominators, pitcher A has 4 of the 7 total cubes or $\frac{4}{7} = \frac{20}{35}$ cranberry; B has $\frac{3}{5}$ of its total cubes or $\frac{3}{5} = \frac{21}{35}$ cranberry. So B has a stronger cranberry taste.

   d. Using common denominators to compare ratios within each pitcher, $\frac{3}{2} = \frac{9}{6} > \frac{4}{3} = \frac{8}{6}$, so B has a stronger cranberry taste.

16. In a true enlargement, all dimensions grow by the same factor. Suppose the original picture measured 5 cm × 6.5 cm. If the picture were enlarged and the width increased from 5 to 9, it grew to 1.8 times its original width; the new length should also be 1.8 times the original length, or 11.7 cm. This means that the 9 cm × 10 cm picture is not its enlargement. Check each pair of pictures to find the cases where the length and width were both multiplied by the same factor in going from the smaller to the larger picture. You will find that picture B is an enlargement of picture C. Both dimensions of C are multiplied by 1.25.

17. If you assume that the rats are dumb enough to stand by and wait their turn, that there is some orderly way of assigning the rats to each of the cats, and agree to disregard the foolishness of fractional cats and rats, here are some solutions:

   i. If 6 cats together kill 1 rat in 1 minute, then it would take 12 cats to kill 2 rats in 1 minute. So if the 12 have 50 times as long to do it, they can kill 50 times as many rats (100 rats).

    ii. If 3 cats kill 1 rat in 2 minutes, then 12 cats (4 times as many cats) could kill 4 rats (4 times as many rats) in 2 minutes. The same 12 cats could kill 25 times as many rats if they have 25 times as long to do it. So the could kill 100 rats in 50 minutes.

    iii. 6 cats/1 rat/1 minute
        6 cats/300 rats/300 minutes
        2 cats/100 rats/300 minutes
        12 cats/100 rats/50 minutes

18. When weights are placed farther from the fulcrum, they will exert a greater effect or downward pull. Weights closer to the fulcrum have a lesser effect. A heavier weight closer to the center may be counteracted by a smaller weight placed farther out. How much "tipping" you get depends on the weight you have put on each side and how far along the arm each weight is placed. In A, there are 3 weights on the left side, each 3 units from the fulcrum or $3(3) = 9$ units of pull. On the right side, there are 2 weights, each 4 units from the fulcrum or $2(4) = 8$ units of pull. The balance will tip to the left. In B, you have $1(4) + 2(3) + 1(2) = 12$ units of pull on the left, and $1(4) + 2(2) + 1(1) = 9$ units of pull on the right. Again, the beam will tip to the left. Also notice that in B, the far weights balance, while the closer weights are farther to the left.

19. A picture may be the best way to represent this situation. Shade $\frac{2}{3}$ of a rectangle to represent the married men and $\frac{3}{4}$ of another rectangle to represent the married women. Because the corresponding numbers of men and women are equal, position the rectangles so that the shaded parts overlap.

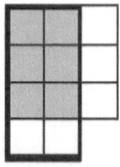

Then you can clearly see that the total number of women is the same as $\frac{8}{9}$ of the total number of men. The ratio of men to women is $9 : 8$.

20. Let A be the coffee sold for \$8 per pound, and B, the coffee sold for \$14 per pound. If we buy A alone, we pay \$8 per pound. If equal amounts of each type of coffee are used in the mixture (25 lb. of A and 25 lb. of B), then we will pay \$11 per pound. There must be more of A because the mixture is selling for \$10 per pound.

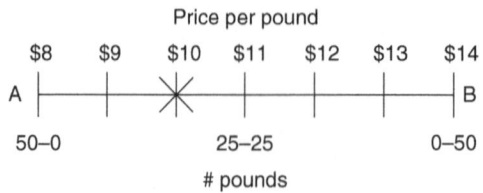

Because the cost of the mixture is $\frac{2}{3}$ of the way between \$8 and \$11, the amount of A must be $\frac{1}{3}$ of the way between 25 and 50 pounds. There are $33\frac{1}{3}$ pounds of coffee A, and the rest of the mixture $\left(16\frac{2}{3} \text{ pounds}\right)$ must be coffee B.

CHAPTER ACTIVITIES

1. Using Mr. Short's measurements, we can see that 2 buttons are as tall as 3 paperclips. If Mr. Tall measures 6 buttons, that would be the same as 9 paperclips.

2. If 2 buttons are as tall as 3 paperclips, then five times as many buttons are as tall as 5 times as many paper clips. This means that 10 buttons have the same height as 15 paperclips. The length of Mr. Tall's car is 10 buttons. If 2 buttons are as tall as 3 paperclips, then 1 button is as tall as $1\frac{1}{2}$ paperclips. $7\frac{1}{2}$ paperclips are as wide as $2 + 2 + 1 = 5$ buttons.

3. In A, notice that 2 out of every 3 eggs are brown. Color B and C accordingly. B should have 8 brown eggs; C should have 12 brown eggs.

4. a. No. 12 ounces should cost \$0.98 and 15 ounces should be \$0.245 more than 12 ounces.
   b. No. 10 pounds should cost twice as much as 5 pounds.
   c. Yes
   d. No. 50 sheets should cost \$3.38.
   e. Yes
   f. No. Buying 1 box is costs less than the price per box when buying a bundle of 3 boxes.

5. Because of his visual representation of the problem, Josh can tell that 2 buttons have the same height as 3 paperclips. $2:3 = 6:9$.

6. Time is not proportional to speed because time decreases as speed increases.

7. If we assume that you can maintain a constant speed for the duration of the trip, then 20 mph × 3 hours = 60 miles, 40 mph × 1.5 hours = 60 miles, and 50 mph × 1.2 hours = 60 miles. Time is inversely proportional to speed and the constant is $k = 60$, the distance.

8. a. $I = \dfrac{k}{d^2}$ b. $F = ka$ c. $F = \dfrac{k}{d^2}$ d. $R = kl$
   e. $m = kr^2$ f. $W = \dfrac{k}{d^2}$

9. Answers will vary.

10. Mo is 1.5 times as along as Louie, so he should get 45 sardines. Pete is half as long as Louie, so he should get 15. You need 90 sardines to feed all three of them.

11. a. ↑↑ and ↓↓   b. ↑↑ and ↓↓   c. ↑↑ and ↓↓   d. NR
    e. ↑↑ and ↓↓   f. ↑↑ and ↓↓   g. ↑↑ and ↓↓   h. NR
    i. ↑↓ and ↓↑   j. ↑↓ and ↓↑   k. ↑↓ and ↓↑   l. NR

12. a, b, c, f, g

13. i, j

14. Variables: number of men working (m), number of days to complete the job (d), number of man days k

    $md = k$

    The number of men working times the number of days they work = the number of man days needed to complete the job.

    The number of men working is inversely proportional to the number of days needed to complete the work.

15. Variables: number of diamonds (d), number of sticks (s)

    constant = k = number of sticks to make one diamond = 4

    $d = 4s$

    The number of diamonds is proportional to the number of sticks.

16. Using the example give, $k = 0.05$. Therefore the speed of the car is given by this equation: $120 = 0.05 \text{ speed}^2$ and the car's speed is just under 49 mph.

17. Charlie is translating at a rate of 5 pages per hour. Working together, they can translate 9 pages in an hour. The constant of proportionality is 350 pages. R × T = 350, so 9 pages per hour = 350. 350 ÷ 9 = 38.88 hours.

18. a. Distance is proportional to time and k = your speed.
    b. Circumference is proportional to diameter and $k = \pi$.
    c. Your cost is proportional to gallons purchased and k = price per gallon.
    d. Pay is proportional to hours worked and k = hourly salary.
    e. Number purchased is proportional to total cost and the unit price is k.
    f. Number of centimeters is proportional to number of meters and $k = 0.01$
    g. Map distance is proportional to real distance and k = scale of the map.

19. Distance and time are proportional. Throughout the trip, $d = kt$. The constant $k = 12$ miles per hour is Jim's average speed.

20. The student's explanations are general and incomplete. The student fails to mention that getting larger or smaller is a multiplicative process, not an additive one. It suggests that the number of apples is proportional to the number of baskets because both numbers are increasing. Suppose you picked 2 apples and your friend picked 10 apples. You could put 2 apples into 1 basket, but if you picked 5 times as many apples (10) you could also put them into one basket. You could pick 10 and put 2 to a basket and use 5 baskets, but this is all speculation. Thus it is not clear that apples and baskets increase by the same factor. The fraction example is better, but also incomplete. It is true that if you multiply a denominator by 3, the value of the fraction is $\frac{1}{3}$ of its original value. For example, changing $\frac{1}{2}$ to $\frac{1}{3}$ multiples the denominator by a factor of $\frac{3}{2}$ $(2 \cdot \frac{3}{2} = 3)$ but, in terms of the fraction values, $\frac{1}{3}$ is $\frac{2}{3}$ of $\frac{1}{2}$. The inverse relationship between the factors operating on the quantities is a critical aspect of inverse proportionality.

SUPPLEMENTARY ACTIVITIES

1.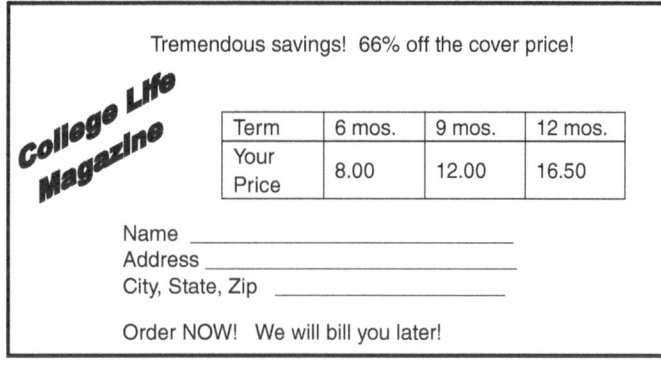

   If we measure Mr. Peterson's height in basketballs, he is 6 balls tall. He is also as tall as a pile of 15 boxes. His son, Matt, is as tall as 4 basketballs. How many boxes will be as tall as Matt?

2. Is there a proportional relationship between subscription term and price? Do you get a better deal if you buy the magazine for a longer period of time?

Tremendous savings! 66% off the cover price!

College Life Magazine

Term	6 mos.	9 mos.	12 mos.
Your Price	8.00	12.00	16.50

Name _____

Address _____

City, State, Zip _____

Order NOW!   We will bill you later!

3. Name the variables and the constant. Write a statement in symbols and in words telling how these quantities are related:

   T = thickness of a book
   N = number of identical books
   H = height of the stack

4. Referring to the book stacking example given above, tell in words what this equation says. $T/H = 1/N$

5. Shade the same proportion of circles in A as in B.

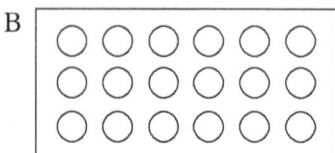

6. Shade the same number of circles in A as in B. What proportion of A is shaded? What proportion of B is shaded?

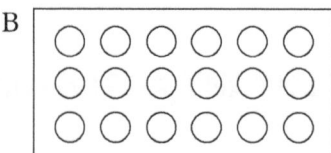

7. Shade the same proportion in B as is shaded in A.

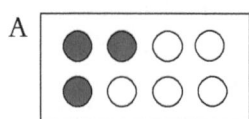
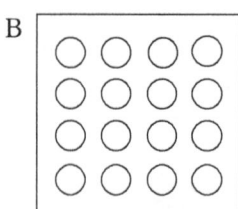

8. In kickboxing, the force needed to break a board varies inversely with the length of the board. If it takes about 5 lbs of pressure to break a board 2 feet long, how many pounds of pressure will it take to break a board that is 6 feet long?

   a. Solve this problem using the constant of proportionality.
   b. Solve by setting up a proportion.

9. Eight people go to a restaurant and 6 are seated at a table, while 2 are seated in a booth. The group orders 3 large pepperoni pizzas and two large artichoke pizzas and agrees that everyone will eat fair shares of both types of pizza. Explain to the waiter how to figure out how much pizza should be delivered to the people in the booth.

10. A middle school student wrote this explanation of direct and inverse proportion. What would you respond?

Direct proportion is like ... my weight is proportional to how much I eat. Like if I eat a lot, I will weigh a lot. But inversely proportional is like my weight is inversely proportional to how much I exercise. If the amount I exercise goes up, then the amount I weigh will go down.

11. Al's peanut butter and Bea's peanut butter are each sold in the sizes listed. The store's flyer advertises the price of Bea's 24 oz. jar as $2.39. You prefer to buy a 64-ounce jar of Al's peanut butter. Assuming that the both brands are priced proportionally, what would you expect to pay for Al's 64-ounce size?

Al's Peanut Butter	Bea's Peanut Butter
8 oz.	6 oz.
16 oz.	24 oz.
64 oz.	48 oz.

12. Construct a situation in which two people lose money on the stock market, in such a way that person A loses a greater proportion of her money than does person B, but loses less money than person B.

13. Jose's typing speed is 55 words per minute and it takes him 1.5 hours to type a handwritten essay for English class. If he composes while he types, he averages about 30 words per minute. About how long does it take him to compose and type the essay?

14. Grant money that was available to people meeting certain eligibility requirements was inversely proportional to the number of people needing the grant. When 30 people needed a grant, they received $250 each. What would have been the amount of each award if 120 people had needed a grant?

PRAXIS QUESTIONS

1. x is proportional to y and the constant of proportionality is 10. If y is positive and the value of y is multiplied by 2, then the value of x is

a. divided by 10   b. multiplied by 10   c. halved
d. doubled   e. impossible to tell

2. If x cannot equal zero and x is inversely proportional to y, which of the following is directly proportional to $\frac{1}{x^2}$?

a. $\frac{-1}{y^2}$ b. y^2 c. $\frac{1}{y}$ d. y e. $\frac{1}{y^2}$

3. x and y are inversely proportional. When $x = 3$, $y = 1\frac{1}{2}$. If x triples, then y

   a. triples

   b. becomes half as great

   c. becomes $\frac{1}{9}$ as great

   d. becomes $\frac{1}{3}$ as great

   e. none of these

4. The village of Grantville reported its population as 10,517 in the 2010 census. The following chart shows how the relative size of the population as reported in the 1930 census. By how much has the population of Grantville grown since the 1930 census?

   1930 ☺☺☺   2010 ☺☺☺☺☺☺☺☺☺☺☺☺☺☺

   a. more than 10,000 people

   b. between 8,000 and 9,000 people

   c. impossible to estimate the growth

   d. about 6,000 people

   e. about 2,400 people

5. If it takes 15 men 20 days to complete a project, how long will it take 2 men to do it?

   a. 300 days   b. 2.66 days   c. 2.66 months

   d. 5 months   e. 1.5 months

6. Twenty-four people can construct a house in 15 days, but the owner would like to finish the work in 12 days. How many more workers should he employ?

   a. 8   b. 6   c. 7   d. 30   e. 3

7. If y varies inversely as x and $x = 6$ when $y = -3$, find y when $x = -9$.

   a. 3   b. −3   c. 6   d. 2   e. −6

8. p varies inversely as q and the constant of proportionality is 20. Find q when $p = 5$; and find p when $q = -2$.

   a. −10 and 5 respectively

   b. −10 and 4 respectively

   c. 4 and −10 respectively

   d. 5 and −10 respectively

   e. 10 and −4 respectively

9. If x and y are inversely proportional and $x = 10$ when $y = 5$, what is y when x is 30?

   a. $\frac{1}{15}$ b. $\frac{5}{3}$ c. $\frac{10}{3}$ d. 4   e. 7

10. How many more awards did person 5 have than person 3? Each 🏆 represents 20 awards.

Person	Number of Awards
1	🏆🏆🏆🏆🏆
2	🏆🏆
3	🏆🏆🏆
4	🏆🏆🏆🏆
5	🏆🏆🏆🏆🏆

    a. 20   b. 40   c. 60   d. 80   e. 100

11. a is directly proportional to b and the proportionality constant is $\frac{2}{3}$, so 15a =

    a. 3b   b. 10b   c. 6b   d. 9b   e. 15b

12. The product of two numbers is 900. One number is tripled. In order for the product to remain the same, the other number must be

    a. multiplied by 3   b. divided by $\frac{1}{3}$

    c. multiplied by $\frac{1}{3}$ d. subtracted from 900

    e. divided by 3 and subtracted from 900

13. If we double the value of both a and c in the fraction $\frac{ab}{c}$, the value of the fraction is

    a. doubled   b. tripled   c. multiplied by 4

    d. halved   e. unchanged

14. Six tractors can plow a field in 8 hours if they all work together. How many hours will it take 4 tractors to do the job?

    a. 9   b. 10   c. 11   d. 12   e. 14

15. The number (n) of 2-inch wooden cubes in a stack is proportional to the height of the stack (h). What is the constant of proportionality?

    a. 2   b. 1   c. $\frac{1}{2}$ d. n·h   e. none of these

CHAPTER 2

Fractions and Rational Numbers

DISCUSSION OF ACTIVITIES

1. One example is a combination problem: Joe's ice cream store sells three different flavors of ice cream, a choice or large or small cone or dish, and four different toppings. How many different purchases can you make at Joe's? A second example is an area problem: A rectangle is 5 cm long and 3 cm wide. What is its area? Another common type is a multiplicative comparison problem: Last month, Jack collected money from 135 customers on his paper route. This month, he collected from $\frac{3}{5}$ as many as last month. How many customers paid their bills this month?

2. A common example is a speed problem: If Mrs. Jones traveled 145 miles in $2\frac{1}{2}$ hours, what was her average speed for the trip? When you divide miles by hours, you get miles per hour.

3. Here is an example: For a project he is doing, Mr. Carter needs 15 pieces of rope each $\frac{3}{8}$ of a meter long. He has 5 meters of rope in the garage. How many pieces can he cut from the rope he has before he needs to buy more?

4. Here is one example: If a yard of materials costs $3.95, how much will $\frac{3}{8}$ of a yard cost?

5. Although you see three different models, a set model, an area model, and a number line model, all three figures represent the same relative amount, $\frac{2}{3}$. In each case, 2 out of 3 equal parts are shown.

6. a. 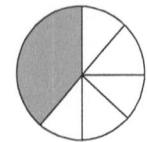   b. 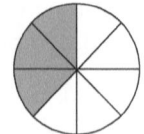   c.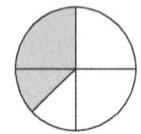

7. Zero is a digit. We use 10 digits (0, 1, 2, 3, 4, 5, 6, 7, 8, 9) in our number system. In this sense, 0 is a symbol. Zero is also a numeral. A numeral is a combination of

digits used to represent a number. We use 0 to represent, for example, the number thirty, 30. Zero is also a number in its own right. It is the next whole number smaller than 1.

8. a. Positive wholes = {1, 2, 3, 4, ...}
    b. Non-negative integers = {0, 1, 2, 3, 4, 5, ...}
    c. Non-positive integers = {..., −4, −3, −2, −1, 0}
    d. Even wholes = {0, 2, 4, 6, 8, ...}

9. Show each can be written in the form $\frac{a}{b}$, where $b \neq 0$.

    a. $-6 = \dfrac{-6}{1} = \dfrac{6}{-1} = -\dfrac{6}{1}$ b. $0 = \dfrac{0}{1} = \dfrac{0}{2}$ c. $2\dfrac{1}{3} = \dfrac{7}{3}$

    d. $0.68 = \dfrac{68}{100}$ e. $54 = \dfrac{54}{1}$

10. a. 3.001   b. 2.304   c. 0.0508   d. 1.1002

11. We call fractions *equivalent* because they are not identical in their appearance, but we are justified in using the symbol = because they are equal in value. They represent the same underlying rational number. So, we can think of the adjective *equivalent* as meaning the same, where the sameness is in value, not appearance.

12. Many people do not realize that mph is a ratio. It really means miles per hour or $\dfrac{mi}{hr}$ and instruction needs to address this explicitly. Technically, we should write mi:hr, but in everyday use this notation is rare, so the fraction notation seems the best way to emphasize that mph is really a comparison of two quantities.

13. The *same relative amount* means that in both pictures, the same portion of the whole is colored. Although the whole amount if different in each picture, the same portion of the whole is colored in each case.

14. Some examples are density, speed, percent, slope.

15. a. Yes   b. No. $\dfrac{-3}{4}$ is not a fraction.   c. Yes
    d. Yes   e. Yes   f. Yes
    g. Yes   h. Yes   i. Yes
    j. No. $1\dfrac{3}{4} = \dfrac{7}{4}$ is not an integer.   k. Yes   l. Yes
    m. Yes   n. Yes

16. b, c, d, e, g

17. a. 15, 0, 564   b. all except $\sqrt{8}$ c. 15, −6, 0, 564, −32
    d. 15, 564

18.   a. $\dfrac{16\,\text{mi}}{1\,\text{hr}}$ b. $\dfrac{16\,\text{mi}}{1\,\text{hr}}$ c. $\dfrac{57.5\,\text{mi}}{1\,\text{hr}}$ d. $\dfrac{84\,\text{mi}}{1\,\text{hr}}$

   e. $\dfrac{6.81\,\text{mi}}{1\,\text{hr}}$

19.   a. $2\,(100) + 4\,(1) + 5\left(\dfrac{1}{100}\text{ of a whole}\right)$

   b. $3\,(1) + 6\left(\dfrac{1}{10000}\text{ of a whole}\right)$

   c. $3\left(\dfrac{1}{10}\text{ of a whole}\right) + 5\left(\dfrac{1}{1000}\text{ of a whole}\right) + 1\left(\dfrac{1}{10000}\text{ of a whole}\right)$

   d. $1\,(1000) + 4\,(10) + 5\,(1) + 6\left(\dfrac{1}{100}\text{ of a whole}\right) + 2\left(\dfrac{1}{10000}\text{ of a whole}\right)$

20.   Children are confused by the lengths of the numbers. A is a very small number although it takes many more numerals to write it. This reflects an incomplete knowledge of place value. In a, 1 represents $\dfrac{1}{10000000}$, while in b, 1 represents $\dfrac{1}{100}$.

21.   a. 18   b. $\dfrac{2}{5}$ c. 3:2   d. $\dfrac{3}{5}$

SUPPLEMENTARY ACTIVITIES

1.   a. Draw 2 $\left(\dfrac{1}{3}\text{-pizzas}\right)$ b. Draw 1 $\left(\dfrac{2}{3}\text{-pizza}\right)$ c. Draw 1 $\left(\dfrac{3}{4}\text{-pizza}\right)$

   d. Draw 3 $\left(\dfrac{1}{2}\text{-pizzas}\right)$ e. Draw $\dfrac{1}{4}$ of 3 pizzas

2.   Fill in the blanks. Do not do any computing. Do this in your head! Use properties, values of decimal fractions, and other hints provided.

   a. $34.8 \times$ _____ $= 0.348$

   b. $0.65 \times$ _____ $= 650$

   c. $0.00234 \times$ _____ $= 23.4$

   d. $\dfrac{3}{8} \times$ _____ $= 1$

   e. $\dfrac{3}{8} \times$ _____ $= \dfrac{300}{32}$ $\left(3 \times ? = 300;\ 8 \times ? = 32;\ \text{get } \dfrac{100}{4} = 25\right)$

   f. $\dfrac{3}{8} \times$ _____ $= \dfrac{35}{72}$

   g. $\dfrac{1}{10000} \times$ _____ $= 2.3456$

   h. $\dfrac{10}{7} \times$ _____ $= 1\dfrac{3}{7}$

i. $\frac{2}{5} \div 4 =$ _____ (*Hint*: dividing by 4 is the same as multiplying the denominator by 4.)

j. $\frac{1}{6} \div 2 =$ _____

k. $0.00384 \times 1000 =$ _____

l. $\frac{4}{7} \times \frac{1}{8} =$ _____

m. $\frac{4}{7} \div \frac{1}{8}$ _____

n. $\frac{4}{7} \div 8 =$ _____

o. $1\frac{2}{3} \times$ _____ $= 1$

p. $\frac{5}{9} \div 5 =$ _____

q. $\frac{2}{3} \times$ _____ $= \frac{24}{18}$

3. Multiplying a fraction by which of these will produce a larger fraction?
   a. $\frac{3}{8}$ b. $1\frac{1}{4}$ c. $\frac{3}{2}$ d. 0.08 e. 1.6 f. $\frac{4}{7}$ g. 1.0004

4. Multiplying a fraction by which of these will produce a smaller fraction?
   a. $\frac{3}{8}$ b. $1\frac{1}{4}$ c. $\frac{3}{2}$ d. 0.08 e. 1.6 f. $\frac{4}{7}$ g. 1.0004

5. Dividing a fraction by which of these will produce a smaller fraction?
   a. $\frac{3}{8}$ b. $1\frac{1}{4}$ c. $\frac{3}{2}$ d. 0.08 e. 1.6 f. $\frac{4}{7}$ g. 1.0004

6. Dividing a fraction by which of these will produce a larger fraction?
   a. $\frac{3}{8}$ b. $1\frac{1}{4}$ c. $\frac{3}{2}$ d. 0.08 e. 1.6 f. $\frac{4}{7}$ g. 1.0004

7. Figure out a way to tell (by looking, not by computing!) which multiplications will result in a product larger than $\frac{2}{3}$, and which will result in a product smaller than $\frac{2}{3}$.

   a. $\frac{2}{3} \cdot \frac{2}{3}$ b. $\frac{2}{3} \cdot \frac{3}{2}$ c. $\frac{2}{3} \cdot \frac{3}{5}$ d. $\frac{2}{3} \cdot \frac{7}{8}$ e. $\frac{2}{3} \cdot \frac{3}{4}$ f. $\frac{2}{3} \cdot \frac{5}{3}$

8. In the rational numbers, multiplication distributes over addition. To multiply $3\frac{1}{2} \cdot 2\frac{1}{2}$, first think of $3\frac{1}{2}$ as $3 + \frac{1}{2}$ and think of $2\frac{1}{2}$ as $2 + \frac{1}{2}$.

Think: $(2 + \frac{1}{2})(3 + \frac{1}{2})$ $(2.3) = 6$

$\left(2 \cdot \frac{1}{2}\right) = 1$ So far I have 7.

Then: $(2 + \frac{1}{2})(3 + \frac{1}{2})$ Half of 3 is $1\frac{1}{2}$; now I have $8\frac{1}{2}$

Half of a half is $\frac{1}{4}$; now I have $8\frac{3}{4}$.

Use this technique to do the following multiplications in your head.

a. $1\frac{1}{2} \cdot 3\frac{1}{2}$ b. $2\frac{1}{4} \cdot 4$ c. $3 \cdot 2\frac{1}{8}$ d. $2\frac{2}{3} \cdot 6\frac{1}{2}$

e. $1\frac{1}{2} \cdot 2\frac{1}{4}$

9. Without doing any writing or computing, order these numbers smallest to largest.

i.   a. $1\frac{3}{8}$ b. 0.06   c. 6   d. $\frac{5}{8}$

ii.   a. 2.3   b. $1\frac{5}{8}$ c. 2.1   d. $1\frac{99}{100}$

iii. a. 0.01   b. 0.1   c. 0.001   d. 0.0001

iv.   a. $\frac{13}{14}$ b. 3   c. 3.1   d. 3.333

v.   a. $\frac{1}{10}$ b. $\frac{1}{15}$ c. $\frac{1}{2}$ d. $\frac{1}{7}$

vi.   a. $\frac{58}{100}$ b. 5.8   c. 0.508   d. $\frac{58}{1000}$

vii. a. $1\frac{1}{2}$ b. 0.4   c. $\frac{3}{4}$ d. 0.04

viii. a. $\frac{5}{8}$ b. $\frac{8}{5}$ c. $1\frac{2}{5}$ d. $\frac{5}{3}$

10. The local bakery always sells $1\frac{1}{2}$ times as many chocolate chip cookies as it sells oatmeal cookies. Answer these questions, or write U for unknown.

a. What is the ratio of chocolate chip to oatmeal cookies sold?
b. What fraction of their daily sales is oatmeal cookies?
c. If the bakery sells 120 chocolate chip cookies, how many oatmeal cookies does it sell?
d. If the bakery sells 300 cookies a day, what fraction of the cookies sold are chocolate chip?

PRAXIS QUESTIONS

1. Triangles form what fraction of this set of figures?

    a. $\dfrac{6}{6}$ b. $\dfrac{1}{2}$ c. $\dfrac{1}{3}$ d. $\dfrac{2}{1}$ e. not given

2. In the fraction $\dfrac{1}{x}$, x could be replaced by all of the following except

    a. 1   b. 0   c. 10   d. 4.2   e. 9

3. Which of the following whole numbers corresponds to the fraction $\dfrac{16}{2}$?

    a. 2   b. 4   c. 6   d. 8   e. none of these

4. Frank has 2 dogs and 5 cats. What fraction of these animals are cats?

    a. $\dfrac{2}{5}$ b. $\dfrac{2}{7}$ c. $\dfrac{5}{2}$ d. $\dfrac{5}{7}$ e. $\dfrac{7}{2}$

5. $\dfrac{1}{100} + \dfrac{1}{1000}$

    a. 0.110   b. 0.011   c. 0.0101   d. 0.01   e. not given

6. If a train covers 14 miles in 10 minutes, then the rate of the train in miles per hour is

    a. 140   b. 112   c. 84   d. 100   e. 98

7. Of the following, the number that is nearest in value to 4 is

    a. 3.985   b. 4.005   c. 4.01   d. 4.1   e. 4.105

8. A micromillimeter is defined as 1 one-millionth of a millimeter. A length of 14 micromillimeters may be represented by

    a. 0.00014 millimeters   b. 0.0000014 millimeters
    c. 0.0014 millimeters   d. 0.014 millimeters
    e. none of the above

9. Choose the correct sum: 0.25 + 0.35 = ?

    a. $\dfrac{7}{10}$ b. $\dfrac{3}{5}$ c. $\dfrac{2}{4}$ d. $\dfrac{1}{3}$ e. none of these

10. All of the following are rational numbers except

    a. 0.0034   b. $1\frac{7}{10}$ c. 0   d. 1   e. π

11. In a school of 800 pupils, 300 are boys. The ratio of the number of boys to the number of girls is

    a. 3 to 8   b. 5 to 8   c. 3 to 11   d. 5 to 3   e. 3 to 5

12. $0.88 for 44 grams is equivalent to

    a. $0.02 per gram   b. $0.01 for 2 grams
    c. 88 grams for $0.44   d. 22 grams for $0.11
    e. $0.22 for 4 grams

13. Choose the decimal whose value is smallest.

    a. 0.0101   b. 0.10001   c. 1.0000001   d. 0.00101   e. 1.00001

14. 0.34 is how many times 34,000?

    a. 0.1   b. 0.01   c. 0.001   d. 0.0001   e. 0.00001

15. In which set of numbers would you find the additive inverse of $\frac{3}{4}$?

    a. Whole numbers   b. Counting numbers   c. Rational numbers
    d. Integers   e. Fractions

## CHAPTER 3

Relative Thinking and Measurement

DISCUSSION OF ACTIVITIES

1. Both the King family and the Jones family have two girls. However, the families have different numbers of children, so that if we think about the number of girls as compared to (or relative to) the number of boys or to the total number of children, then the situation looks different in each family. In the Jones family, $\frac{2}{5}$ of the children are girls and in the King family, $\frac{2}{4}$ or $\frac{1}{2}$ of the children are girls. The ratio of girls to boys in the Jones family is $2:3$, while in the King family, it is $2:2$. Therefore, the King family has a greater proportion of girls.

2. Merely by counting the brown eggs in each container, we can tell that the 18-egg container has more brown eggs. However, we might consider the fact that the containers hold different numbers of eggs. Seven out of the 18 eggs or $\frac{7}{18}$ are brown. Six out of the 12 eggs or $\frac{6}{12}$ or $\frac{1}{2}$ of the dozen are brown. Seven is less than half of 18, so a greater portion of the dozen eggs is brown.

3. Yes, Bert is correct.

4. The answer to "how much" is not a number of slices. Pan B contains $\frac{3}{8}$ of a pizza more than pan A contains. You could serve up the amount of pizza in pan A $2\frac{1}{2}$ times from the pizza in B. Note that these comparisons are OK in this situation because the pizzas are the same size. If they were not, these questions would not be meaningful.

5. You can tell who walked farther merely by adding the distances that each person walked. Dan walked 6 miles and Tasha walked 7 miles. To decide who walked faster, you need to compare distance to the time it took to walk it. Dan traveled his 6 miles in 2.5 hours, so his speed was $\frac{6\,\text{mi}}{2.5\,\text{hr}} = 2.4\,\frac{\text{mi}}{\text{hr}}$. Tasha walked her 7 miles in the same amount of time, so her speed was $\frac{7\,\text{mi}}{2.5\,\text{hr}} = 2.8\,\frac{\text{mi}}{\text{hr}}$. Tasha walked faster.

6. It is more helpful to know the percent of discount because then you know the amount you can save on an item of any price. 20% off means that you will save $0.20 for every dollar that you spend. If someone tells you only a dollar amount, you cannot tell if it is a good sale or not. $2.00 off on an item whose original cost was $500 is hardly worth the effort of rushing down to the store, but a saving of $2.00 on a $5.00 item is a substantial saving (40%).

7. To judge the crowdedness of an elevator, you need to know how many people are on the elevator; however, that is not enough information. You might not call it crowded if there are ten people on an elevator that holds 25 people, but if there are 10 people on an elevator that holds 8, things are tighter! The number of people on the elevator must be compared to the recommended capacity or to the floor area (the number of square feet of floor space) in order to be able to tell if the elevator is crowded or not. If both elevators have the same capacity, the choice is clearly A, but if they do not have the same capacity, relative thinking is needed.

8. At table A, for each root beer served, $1\frac{1}{2}$ colas were served. At table B, for each root beer served, $1\frac{1}{3}$ colas were served. You can think of it this way: At table B, the root beer is not as diluted by cola as the root beer at Table A, so the people at table B consumed more root beer.

9. Leesa is correct.

10. Some possibilities are:
   a. What part of the pack did the girls chew? Is a greater portion of the package chewed or saved for later?
   b. How many cans of pop remained? What part of the 6-pack was given to the food pantry?
   c. How many pieces did they eat? What part of the pizza did they take home?
   d. How many days did Thomas work? What part of a month did he not work?
   e. How many of the eggs were white? What part of a dozen is brown?

11. The first three responses are additive. The children think that if you add more people, you should buy more pounds of peanuts. Unfortunately, when children like Grant think they see a pattern, they are sometimes convinced that their response is correct. Mark's response was the most productive because he realized that he should compare number of people to number of pounds. He was not able to break down his solution to a half pound for 2 people, but he knew that he would have extra by buying 3 pounds.

12. How sour a mixture is depends not only on the amount of vinegar, but on the amount of vinegar as compared to the amount of sugar. We might compare the number of teaspoons of vinegar to the number of teaspoons of sugar:

$$\frac{\#\ \text{teaspoons of vinegar}}{\#\ \text{teaspoons of sugar}}$$

Using this ratio, the higher number would indicate the more sour taste. For bowl A, $\frac{8}{10} = 0.8$ and for bowl B, $\frac{6}{8} = 0.75$.

13. What does it mean to be square? In a square, two adjacent sides have the same length, so if you compared them, you would get a ratio of $1:1$. Find the ratios of adjacent sides in each of the rectangles and see how close they are to 1. The ratio of adjacent sides for the $114' \times 99'$ rectangle is $\frac{99}{114}$ or 0.87. The ratio of adjacent sides for the $455' \times 494'$ rectangle is $\frac{455}{494}$ or 0.92. The ratio for the $284' \times 265'$ rectangle is $\frac{265}{284}$ or 0.93. The $284' \times 265'$ rectangle is most square.

14. Ty doesn't understand that inverse relationship between the number of parts and the size of the parts. You might get him to think about it by asking, "Your mom has 2 identical cherry pies. She cuts one into 9 equal pieces and the other into 5 equal pieces. You really like her cherry pie. From which pie do you want to have your piece?"

15. Fred does not understand the compensatory principle of measurement. The size of the pizza (its area) isn't going to change. If you cut it into more slices, each slice will be smaller; if you cut it into fewer slices, each slice will be larger.

16. The blackness of the ink depends on the amount of amount of water added. Both companies use a recipe in which the water added is half as much as the amount of black dye, so we would expect both ink mixtures to look equally black. To measure blackness of the ink, we need to compare the amount of black dye to the amount of water. For the India Ink company, $\frac{\#\ parts\ black\ dye}{\#\ parts\ H_2O} = \frac{3}{1.5} = 2$. For the Midnight Ink Company, $\frac{\#\ parts\ black\ dye}{\#\ parts\ H_2O} = \frac{2}{1} = 2$. Our measure confirms our expectations.

17. Assuming you like cookies, *better* means that if you joined the group, you would get more cookies. If you jointed Tom's group, there would be 8 cookies for 5 people, or $\frac{8}{5} = 1.6$ cookies for each person. If you joined Jenny's group, there would be 11 cookies for 6 people, or $\frac{11}{6} = 1.8$ cookies per person.

18. The steepness of a ramp is going to depend on the horizontal distance over which that rise occurs. If it occurs in a very short distance, the ramp will be steep; if it occurs over a long distance, the ramp will not be too steep. If you made the ramp to the doorway that is $5'$ high the same length as the one to the $2'$ high doorway, then the ramp to the $5'$ high doorway would be very steep. However, if you started that ramp farther away from the building, you could decrease its steepness.

By starting sufficiently far away from the building, you could build it of the same steepness as the ramp to the 2′ high door.

19. By considering only the amount of food eaten, clearly, the animals can be ranked: b, e, a, c, f, d. (Don't forget to use the same units of measure when comparing!) However, if you consider amount eaten relative to body weight, you get a different (counterintuitive) result. The smallest to largest eaters are: c, d, a, f, b, e.

20.   a.   10 inches   b. $9\frac{1}{2}$ inches   c. $9\frac{3}{4}$ inches   d. $9\frac{5}{8}$ inches

21.   but not . Whether you are partitioning a unit of length or a unit of area, you must be sure that you make pieces of equal length or of equal area. If you partition the triangle using horizontal lines, you cannot be sure that your pieces are of equal area.

22. B has $1\frac{1}{2}$ measures of sugar for each part of red juice. In A, every part of red is matched with $1\frac{1}{3}$ measures of sugar. Mixture B is sweeter because $1\frac{1}{2} > 1\frac{1}{3}$.

23. In A, each cup of concentrate is diluted by $1\frac{1}{2}$ c. water. In B, each cup of concentrate is diluted by 4 c. water. In C, each cup of concentrate is diluted by $1\frac{2}{3}$ c. water. In D, each cup of concentrate is diluted by $1\frac{3}{4}$ c water. So B, having the most water per concentrate, is weakest. Next is D, then C. A is the most orangey.

24. Mary's solution is a counting solution. She devised a way to make counting the chips a little easier, but she does not acknowledge that the size of the cookies might be important. Instead of trying to count all of the little chocolate pieces, she assumes that the chocolate chips are evenly distributed throughout, and decides to count the chips in 1 of 8 equal-sized pieces, then multiply by 8. Her solution says that the one with the most pieces of chocolate, regardless of size, will be chocolatier.

   Both Jim and Tony realize that merely counting chips will not measure how chocolaty a cookie is, and that the size of the cookie is important. Jim suggests using the area of each cookie and the number of chocolate pieces in each cookie. However, he does not convince us that he realizes that the number of chocolate chips must increase proportionately with the number of square inches. It is not clear whether he is using the word *more* in a proportional sense or an additive sense. Would he say that if the smaller cookie had 49 chocolate chips and the larger cookie had 51, the larger cookie would be chocolatier? I would ask Jim to explain. Tony takes a slightly different approach. He suggests that we take the same size piece of each cookie—that is, hold the area constant—and count the chocolate chips in the bite from each cookie. He makes the assumption that the chocolate chips are evenly distributed. Like Jim, he wants to compare number of chocolate chips to an area. The difference is that he suggests equalizing the area from each cookie, but then the task is reduced to comparing the numbers of chocolate chips. Although his *bite* is not very accurate measure of area, this is a pretty good strategy.

25. ET got the correct answer for the wrong reason. When he had 1Y in A and 2Y in B, he jumped to the conclusion that recipe A was more orangey. However, he failed to realize that the Y in mixture A had to be compared to 2O, and the 2Y in B had to be compared to 3O. By his reasoning, if recipe B had been 3O : 4Y, the mixtures would have the same! MS did a purely subtractive process and never engaged in any comparative thinking. She ended up with 1O : 2Y in mixture B and should have compared with the 2O : 3Y in mixture 1. The third student gave an incomplete explanation. What he said is true; B would be lemony. However, he stopped short of checking whether B is more or less lemony than the recipe in A.

SUPPLEMENTARY ACTIVITIES

1. A    B

    a. How many more stars are there in box B than in box A?
    b. How many times the number of stars in A are in B?
    c. Set A is what part of set B?
    d. If there are two stars in a group, how many groups did it take to make Box A? Box B?

2. Can you tell by looking at the following pools which one has a greater capacity? Which pool has the greater area? In addition to the area, what quantity is needed to describe capacity? By choosing appropriate depth for each of the pools, show that sometimes the first pool might have the greater capacity, and sometimes the second might have the greater capacity.

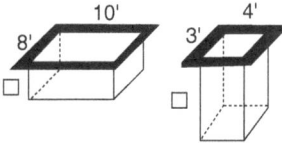

3. Today there is a sale at Neville's Women's Shop: $10.00 off the sticker price of every item! Describe exactly how you can tell which is a better buy, a dress marked $44.99 or a blouse marked $26.99.

4. What does it mean to say that a grasshopper can jump relatively higher than a person can?

5. Who has the larger appetite?

eats ◆◆◆◆◆ in 7 days.   eats ▥ ▥ ▥ in 4 days.

6. The Athlete's Foot Stop sells Spikey and Freerock shoes. Last week they sold 100 pairs of Freerocks and 15 pairs of Spikeys. This week, they sold 200 pairs of Spikeys and 45 pairs of Freerocks.

    a. If a student were thinking in absolute terms, explain how he would decide which brand showed the greatest increase in sales this week over last week.
    b. If a student were thinking relatively, explain how she would decide which brand showed the greatest increase in sales this week over last week.

7. The Manhattan Mercury reported that in 1950, when there was a population of 19,000 people, there were 12 people per acre in the city of Manhattan, Kansas. By 1995, when the population was 43,000, there were only 5 people per acre. Interpret this information as completely as possible.

8. There were 7 males and 12 females in the Dew Drop Inn on Monday night. In the Game Room next door, there were 14 males and 24 females. Which spot had more females?

9. The first picture shows Jeb and Sarah Smart when they were younger. The second shows them as they look now. Who grew faster between the first and second pictures?

10. A car was traveling at a speed of 65 mph and a truck was traveling on the same road at a speed of 60 mph. They came to a hill and both vehicles slowed down by 15 mph as they climbed the hill. Which vehicle had a harder time maintaining its speed?

11. Adam, a fifth grader was asked the following question about growing trees. Analyze his response.

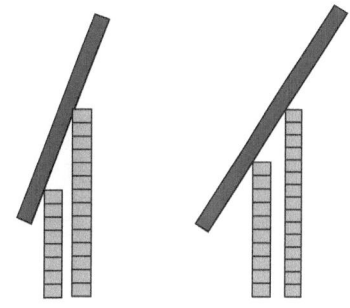

A tree that was 8 feet high last year is 14 feet high this year. Another tree that was 10 feet high last year is also 14 feet high this year. Which tree grew faster?

Adam used unifix cubes to build models of the trees, showing last year's heights and this year's heights. He placed a ruler on top of both models of tree A and another ruler on top of both models of tree B. "The ruler that goes up the most tells me the one that grew faster," he explained.

12. Gordon is making iced tea. The jar says to use $\frac{1}{2}$ c of the iced tea mix in 2 liters of water. Gordon wants to make 6 liters. How much iced tea mix should he use?

13. Here is way some fourth grade students handled question 12. Analyze the strategy each child used. Which strategies are useful? Tell why the others are not appropriate.

Jeanne   It is a pattern. For 2 l you use $\frac{1}{2}$ c so for 6 you would use $\frac{1}{6}$ c

Aaron $\frac{1}{2}$ C   2 l
  6 ÷ 2 = 3
  So I would use
  8 ÷ 2 = 16 C

Chris   double and go in between
  1 c = 4 l
  2 c = 8 l
  1½ c = 6 l

Lee $\boxed{\frac{1}{2}}$ $\boxed{\frac{1}{2}}$ $\boxed{\frac{1}{2}}$ Use 3 half c
  2 l   2 l   2 l

Pat   Increase everything by 4
  2 l + 4 l = 6 l
  $\frac{1}{2}$ c + 4 c = $4\frac{1}{2}$ c

14. Below you see a strip placed on top of a ruler.

   a. What is its length to the nearest inch?
   b. To the nearest half inch?

c. To the nearest quarter inch?

d. What is its length to the nearest sixteenth of an inch?

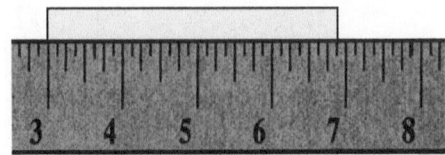

15. Which property is more square?

Property A: 60′ × 90′   Property B: 75′ × 110′

16. Here is Sen's response to question 15. Is he correct?

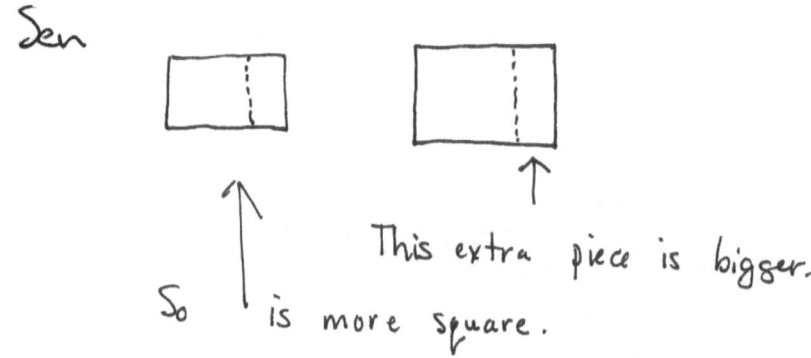

Sen

This extra piece is bigger.

So [is more square.

17. How steep is this ramp?

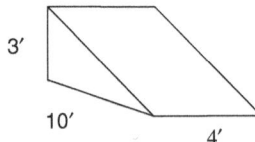

3′

10′   4′

18. Here is Mark's response to problem 17. Is he correct?

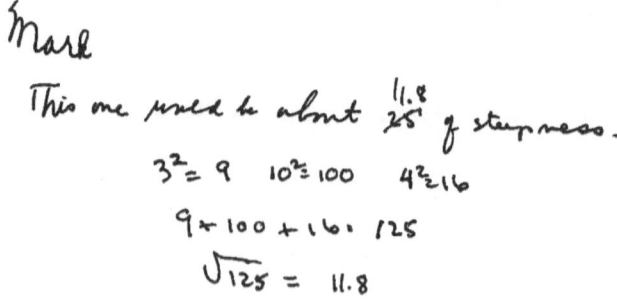

Mark

This one would be about $\frac{11.8}{25}$ of steepness.

$3^2 = 9 \quad 10^2 = 100 \quad 4^2 = 16$

$9 + 100 + 16 = 125$

$\sqrt{125} = 11.8$

19. Sam and Jason, two third graders, commented on the following pictures:

Sam said that $\frac{7}{7}$ is larger because there are more pieces. Jason said that $\frac{4}{4}$ is larger because the pieces are bigger. How would you explain this situation? Which measurement principle might the boys not understand?

20. I saw this sign when I was driving in the mountains. What does it mean?

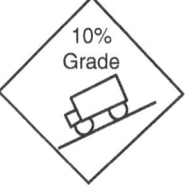

21. Which of these figures is rounder? How do you measure roundness?

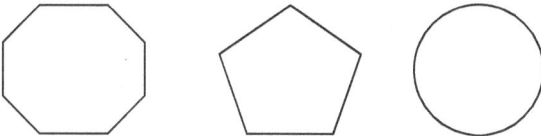

22. The Chicago Bulls and the Detroit Pistons once played each another in January and in March of the same season. The games brought two great stars into the same court. In January, Michael Jordan scored 30 points and Isaiah Thomas scored 20 points. In March, Jordan scored 40 points and Thomas scored 30 points. The following bar graph shows this information. Explain which player's performance improved the most.

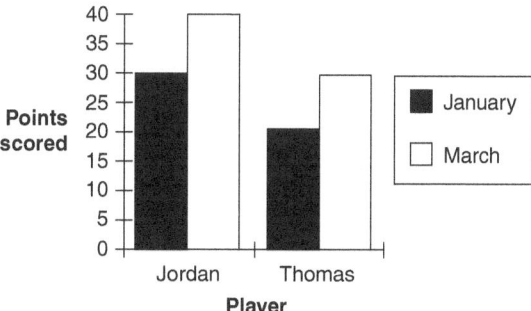

23. Revisiting the Bland Company: Which formula will be darker brown?

Formula A   Formula B

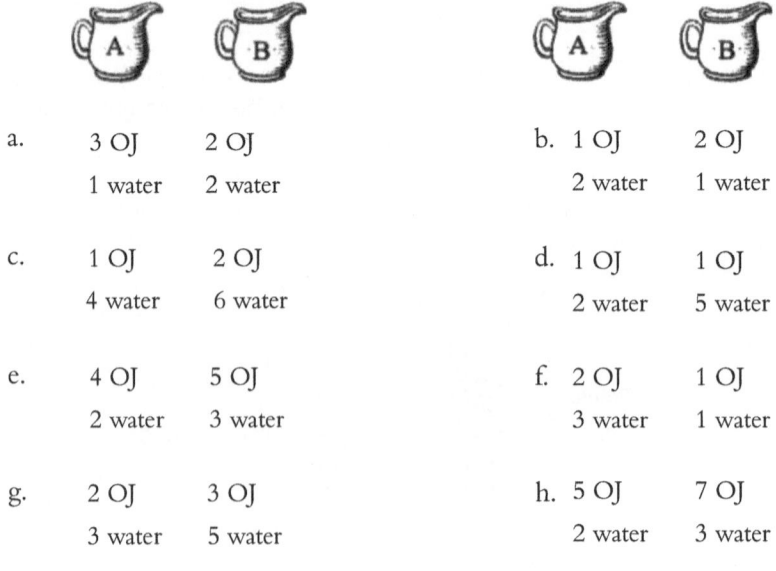

24. You have 2 pitchers, A and B, and in each of them you mix up a recipe of orange juice concentrate and water. For each pair of recipes below, decide which is going to have the more orangey taste and be able to support your answer.

	A	B			A	B
a.	3 OJ 1 water	2 OJ 2 water		b.	1 OJ 2 water	2 OJ 1 water
c.	1 OJ 4 water	2 OJ 6 water		d.	1 OJ 2 water	1 OJ 5 water
e.	4 OJ 2 water	5 OJ 3 water		f.	2 OJ 3 water	1 OJ 1 water
g.	2 OJ 3 water	3 OJ 5 water		h.	5 OJ 2 water	7 OJ 3 water

i.   2 OJ   3 OJ   j.   1 OJ   2 OJ

    3 water   4 water   2 water   4 water

25. A fifth grade class conducted a lemonade sale. They sold 8 oz. cups of lemonade for $0.40 each. In 30 minutes, they sold 15 cups for a total of $6.00. The next day in math class, each group of 3 children were given a cup with a black line marking 2, 4, or 6 ounces. They were asked how much they would charge for the amount of lemonade marked on their cup and how many cups of that size they would have to sell to make $6.00.

   Analyze the reports from groups A, B, C, and D, and try to understand the strategy each group used. What does each report suggest about the students' (a) comfort with multiplication and division; (b) ability to scale up and down; (c) ability to simultaneously coordinate two sets of numbers (ounces and prices)?

Group A: How many 4-ounce cups do you need to sell? What will be your price?

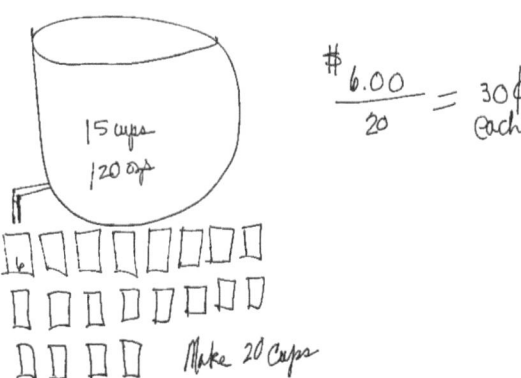

Group B: How many 6-ounce cups do you need to sell? What will be your price?

Group C: How many 12-ounce cups do you need to sell? What will be your price?

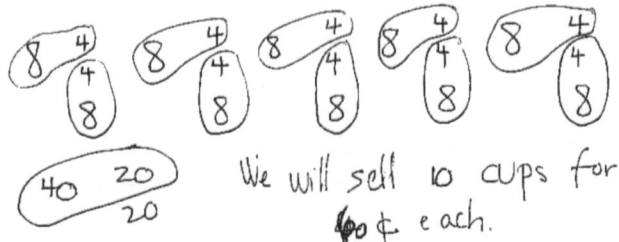

Group D: How many 6-ounce cups do you need to sell? What will your price be?

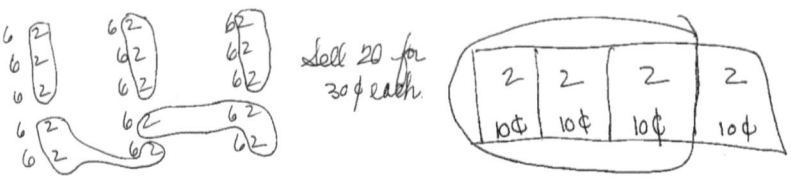

PRAXIS QUESTIONS

1. The ratio of boys to girls in a particular classroom is $6:5$. How many boys are in the class?

    a. $\dfrac{5}{11}$　　　b. $\dfrac{6}{11}$　　　c. 12　　　d. 11　　　e. can't tell

2. A ribbon measures $6\dfrac{15}{16}$ inches long. What is its length to the nearest one-quarter inch?

    a. 7　　　b. $6\dfrac{7}{8}$　　　c. $6\dfrac{1}{2}$　　　d. $6\dfrac{3}{4}$　　　e. $6\dfrac{5}{8}$

3. Which of the following measures could not be the height of a real building?

    a. 0.3 kilometers　　　b. 60,000 millimeters　　　c. 23 centimeters
    d. 29 meters　　　e. 3.5 kilometers

4. Pizza A has a 12-inch diameter and sells for $12. Pizza B has a 14-inch diameter. If your unit pricing is the same no matter what the size of the pizza, how much should you charge for pizza B?

    a. $14　　　　　　b. increase price of A by $\dfrac{1}{6}$ or 16.7%
    c. 36% more than $12　　d. $1.50/slice　　　　　　e. $2/slice

5. Which of the following mixtures of orange juice concentrate (O) and water (W) will have the weakest orange flavor?

    a. 2O, 3W   b. 3O, 4W   c. 5O, 7W   d. 8O, 9W   e. 4O, 5W

6. You mixed 5 tablespoons (T) of juice mix with 4 cups (c) of water, and then discovered that the suggested recipe was 3T juice mix to 2 cups of water. What can you add to fix your drink?

    a. $\frac{1}{4}$ c. water   b. 1 T juice mix   c. $\frac{1}{4}$ c. juice mix

    d. 2 T juice mix   e. can't be fixed

7. Which of the following could be the length of a sofa?

    a. 75 cm   b. 4 meters   c. 150 mm

    d. 1.2 decimeters   e. 0.5 kilometers

8. If batch A of iced tea was mixed with 5 parts tea and 4 parts water and batch B was mixed with 2 parts tea and 3 parts water, then

    a. B would taste stronger because it has less water.
    b. A would taste stronger because it has more tea.
    c. A would be stronger because it has 1 part tea that is not matched by a part of water.
    d. They would taste the same because each has one more part water.
    e. none of the above

9. Which of the following quantities cannot be measured directly?

    a. mass   b. height   c. time   d. slope   e. all of the above

10. A pedometer converts all distances to meters. The display for a distance of 0.5 kilometers would be 0.5 multiplied by

    a. 100   b. 2   c. $\frac{1}{2}$ d. 1000   e. $\frac{1}{10}$

11. You paid $0.88 for 44 grams. What was the unit price?

    a. 4 grams for $0.02   b. $0.02 per gram   c. half a cent per gram
    d. $0.44 for 88 grams   e. none of the above

12. Which of the following is most unlike the others?

    a. 3% tax increase   b. $3 surcharge   c. 30% discount

    d. 10% tip   e. $\frac{2}{3}$ reduction

13.   Two ladders are leaning against a wall. Which of the following would be a good way to determine the steepest ladder?

    a.   Find out which ladder is longer.
    b.   Measure to see which base is farther from the wall.
    c.   Determine which top rung is higher.
    d.   Determine which one has more distance between its rungs.
    e.   none of the above

14.   Which of these is an absolute quantity?
    a.   elapsed time
    b.   body mass index
    c.   relative humidity
    d.   US poverty level
    e.   inflation rate

15.   How would you find percent increase in price?
    a.   subtract the old price from the new price
    b.   divide the change in price by the old price
    c.   divide the change in price by 100
    d.   divide the change in price by the new price
    e.   divide the new price by 100

CHAPTER 4

Quantities and Covariation

DISCUSSION OF ACTIVITIES

1. In a caricature, the artist deliberately draws some feature(s) out of proportion to the rest of the face or body. Mark Twain's long white hair and mustache give him a kindly, grandfatherly appearance. His ear is way out of proportion with the rest of his head, perhaps suggesting that he would be a good listener. When an artist portrays a political figure, he may be trying to reflect personal or popular sentiment. In the second picture, Prince Charles' ears are drawn out of proportion to the rest of his head, thus giving him a "Dumbo" look.

2. The pizza was not distributed fairly among the three groups. The 45 students received a smaller portion of the pizza than the 45 parents and teachers received, and the parents received a larger portion than the teachers. The portions were not proportional to the numbers of people who ate from them.

3. As always, answers are not necessarily unique.

    picture : frame :: yard : fence   R : is enclosed by
    giraffe : neck :: porcupine : quills   R : is best known for
    food : body :: rain : earth   R : nourishes
    car : gasoline :: sail : wind   R : is fueled by
    sap : tree :: blood : body   R : flows through
    sandwich : boy :: carrot : rabbit   R : is food for
    pear : tree :: potato : ground   R : is grown in
    tree : leaves :: book : pages   R : is composed of
    conductor : train :: captain : ship   R : is the commander of
    wedding : bride :: funeral : corpse   R : is a ceremony for

4. a. No. Orange is twice the length of yellow, but green is more than twice the length of white.
    b. No. Green is 3 times the length of white, but dark green is only twice the length of green.
    c. No. Green is half the length of dark green, but magenta is less than half the length of blue.
    d. Yes. Magenta is half of brown and yellow is half of orange.
    e. No. Blue is 3 times as long as green, but red is only twice as long as white.

5. a. 1 to 1   b. 1 to 4   c. 1 to 2   d. 1 to 3

6. Of course, the answer depends on the height of the person answering the question. Assume that the height of the dog is roughly $\frac{1}{4}$ the height of the person (using knee, waist, chest, top of head as quarter marks) and that the person answering the question is about 5 feet tall. That means the dog would be about $1\frac{1}{4}$ feet high. Then if we increase proportionately, the person grows to $1\frac{3}{5}$ of his or her height, or just under $1\frac{1}{2}$ times his or her height. The dog must do the same. This would make the dog just under 2 feet or about $1\frac{7}{8}$ feet high.

7. a. 2 orange   b. orange + magenta   c. magenta
   d. red   e. black

8. a. Dianne was faster today.
   b. She was slower today.
   c. She was faster today.
   d. She was faster today.
   e. We cannot draw any conclusions if Dianne ran fewer laps in less time than she did yesterday.

9. a. Changing quantities include the amount of gas in your tank, the number of gallons registered on the gas pump, the price registered on the gas pump, and time.
   b. Changing quantities include time, distance from your starting point, amount of gas in your tank, and your speed.
   c. Changing quantities include shopping time, the amount of each purchase, and your total spent.
   d. Changing quantities include time since you started, your depth, pressure, amount of air in your tank, and the amount of sunlight reaching you.
   e. Changing quantities include time, depth of the water in the tub, and volume of water in the tub.

10. a. The amount of money put into the machine changed; time changed; the ratio (2 dimes + 1 nickel) : 1 quarter did not change.
    b. The size of the flower changed, but the shape of the flower did not change.
    c. The real distances between the cities do not change; the distance between any two cities on the map does not change; the ratio of 1 inch on the map to 20 miles in reality does not change.
    d. The length of the room does not change; the unit of measure changes; the relationship between inches and feet does not change (12″ = 1 ft).
    e. The number of pies changes; the cost of each package changes; the cost per pie (or the cost per 7 pies) does not change.

11. a. There are more peas than carrots.
     b. Ted is taller.
     c. More people have cats.
     d. It will cost less than $3.
     e. More fathers than mothers came to school; some students had both parents present.

12. a. $\downarrow\uparrow$ 8 hours
         Here is a way to think about this situation: The 8 people each work for 2 hours; that gives a total of 16 hours of work. If 2 people share the work, they will work together for 8 hours.
     b. $\downarrow\uparrow$ $6
     c. $\downarrow\uparrow$ 12 hours
     d. People's ages do not work that way. Every son does not have a mother 3 times as old as he is.
     e. $\downarrow\uparrow$ 6 hours. Think about this situation the same way we did about part a.
     f. Sales tax doesn't work that way. The percentage stays fixed no matter what size purchase you make.
     g. Increasing the number of people does not increase the amount of time for the meal.
     h. $\downarrow\uparrow$ 10 minutes

13. a. Same number of cookies shared by fewer kids means each gets more.
     b. More cookies shared by the same number of kids means everyone gets more.
     c. More cookies shared by fewer kids means everyone gets more.
     d. Can't tell. You will need another method for comparing these fractions.
     e. Same number of cookies shared by more kids means everyone will get less.
     f. Can't tell. You will need another method for comparing these fractions.
     g. Can't tell. You will need another method for comparing these fractions.

14. What changes? The tables get moved around. What does not change? The number of people at a table and the number of candy bars do not change. Note that these are the two quantities that determine how much each person will get. Where the tables are located is irrelevant. Regardless of the table arrangement, there are still 2 candy bars to be shared by 3 people and everyone will get $\frac{2}{3}$ of a candy bar.

15. What is the same about both orders? Each has 1 whoopie pie and 1 cookie. If you take out what costs the same in both orders, what is left must account for the price difference. So, 1 whoopie pie must cost $3 more than 1 cookie. The first order is like 1 cookie + 1 cookie + 1 cookie + $3, which means that 3 cookies cost $3, or $1 per cookie. Now look at the second order. If 1 cookie costs $1, the 2 whoopie pies cost $8, or $4 each. BTW, do you know why they are called whoopee pies? Whoopie pies are a Pennsylvania Amish tradition. The women packed these treats in their husbands' lunches and when the men opened their lunch pails, they shouted "Whoopie!"

16. Take out the items that appear in both groups, 1 bear and 1 train. These items contribute the same amount to each price. What remains must account for the price change between the two groups. This tells you that a bear cost $4 more than a train. So the second group is equivalent to train + train + train + $4. This means that 3 trains cost $72, and 1 train costs $24. Now you can find that a bear costs $28.

17. What doesn't change? Each group costs $50 and each contains 1 football and 1 helmet. If you take out the like elements in each group, you know that what remains in the first group costs the same as what remains in the second group. So, 1 football costs the same as 2 helmets. Looking at the second group, which is equivalent to 5 helmets, you can see that 1 helmet costs $10 and a football costs $20.

18. Aspect ratio means the ratio of the longer dimension to the shorter dimension of a picture. The aspect ratio of the clip art is $2.51 : 2.44$. If we divide this ratio, we get 1.03. We recognize that this is just a little more than 1. Therefore the picture is nearly square.

19. Look at the longer side of the rectangle as compared to the longer side of the paper. $2 \cdot (5.5) = 11$. Remember that if the rectangle is to be similar, the same scale factor must be applied to both dimensions. So 5.5 is the largest factor possible. If I use a scale factor of 5.5 on the other side of the rectangle, I get $(1.5) \cdot (5.5) = 8.25$. So, the largest similar rectangle will have dimensions 8.25″ by 11″ and the scale factor that gives it is 5.5.

20. If $2\frac{1}{2}'$ is $\frac{1}{20}$ of the actual size, then the height of the actual dinosaur used in the movie was $20 \cdot \left(2\frac{1}{2}\right) = 50'$.

21. The area of A may be calculated like this: Top piece of 2 sq units + bottom piece of 8 sq units = 10 sq units. The area of B is then 8 sq units in the top + 32 sq units in the bottom = 40 sq units. Therefore, the area of A has been increased by a factor of 4 in figure B by doubling the lengths of the sides.

22. a. The scale model of the tennis court should measure 6 cm by 11.5 cm.
    b. The scale model of the sheet of paper should measure 2.5 mm by 4.5 mm.
    c. The scale model of the desktop should measure 17.5 mm by 11 mm.

25. The scale factor is $\frac{1}{2}$.

26. A, C, E, and G are similar. B, H, and I are similar. D and F are similar. J is not similar to any of the others. One way to decide which are similar is to compare the aspect ratios of the figures.

27. If you put the right angle of each triangle at the origin (0,0) with one side along the y-axis and one side along the x-axis, then the third sides of similar triangles should be parallel.

It appears that none of the triangles are similar and this can be confirmed by checking the ratios of corresponding sides.

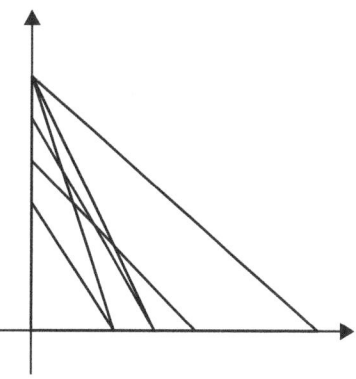

28. a. The side of the smaller square is $\frac{1}{4}$ the side of the larger.

b. The scale factor is $\frac{1}{4}$.

29. a. B is an enlargement of A.

b. The scale factor is $\frac{3}{2}$ because every length on figure A is $1\frac{1}{2}$ times as long on figure B.

c. Right angles were preserved.

d. Vertical segments became $1\frac{1}{2}$ times as long on B as they were on A.

e. Horizontal segments became $1\frac{1}{2}$ times as long on B as they were on A.

f. The area of B is $\frac{9}{4}$ times the area of A.

g. The ratio of corresponding lengths is $3:2$.

h. Shrinking and enlarging preserve angles. Shrinking or enlarging multiplies each length by the scale factor and each area by the scale factor squared.

30. Compare the areas to the prices of the pizzas.

Area (sq. in.)	Price ($)
78.54	6.80
113.1	8.50
153.94	13.60
314.16	28.00

a. Compare price to number of square inches. For the 10″ pizza, you pay $0.086 a square inch; for the 12″ pizza, $0.075 a square inch; for the 14″ pizza, $0.088 a square inch; for the 20″ pizza, $0.089 a square inch. The 12″ pizza is the best buy.

b. No. The pizzas are not priced proportionally. For example, 78.54 sq in doubled is less than 153.94 sq in, but the price for 153.94 sq in is double the price for 78.54 sq in.

31. a. The original mast must have been $(2.5)(14) = 35'$ tall.

 b. The original fish must have been $80\left(\dfrac{1}{25}\right) = 3.2$ feet long.

 c. The original area must have been $(25)\left(\dfrac{1}{4.5}\right)\left(\dfrac{1}{4.5}\right) = 1.235$ sq ft.

32. a. Figures would appear too wide.
 b. Figures would appear slightly taller.
 c. Everything would look ok.

33. a. This means that your distance should be 4.5 times the size of your screen. Don't forget to convert inches to feet. Sit 16 feet from a 32-inch TV.
 b. Sit 22.5 feet from a 60-inch TV.
 c. If your room size will allow you to sit only 16 feet (or 192 inches) away from the TV, then ask yourself "4.5 times what distance gives 192?" The recommended TV size would be 42 or 43 inches.

34. The scale ratio is $\dfrac{4}{1.5} = \dfrac{40}{15} = \dfrac{8}{3}$. This means that a side that measures 8 will become 3. So A′B′ = 8. In the first figure, side CD measures $\dfrac{3}{4}$ of side AB, and after the reduction, C′D′ must measure $\dfrac{3}{4}$ of A′B′ . So C′D′ measures

$$\dfrac{3}{4} \cdot 3 = \dfrac{9}{4} = 2\dfrac{1}{4} = 2.25.$$

35. Notice that ABC is an equilateral triangle (all 3 sides are the same measure), so if AB's length of 5 becomes 4.25, then B′C′ and A′C′ will also measure 4.25.

36. a. The scale ratio is $3 : 10$, so $7.5\left(3 + 3 + 1\dfrac{1}{2}\right)$ inches must represent $(10 + 10 + 5)$ miles or 25 miles. Another way to think about this: $7\dfrac{1}{2}$ is $2\dfrac{1}{2} \cdot 3$, so you need $2\dfrac{1}{2} \cdot 10$.
 b. In your model, 7 inches represent 14 feet, every inch represents 2 feet.
 c. For every 6 miles, you need a map distance of 2 inches, so if you have 24 miles your need a map distance of 8 inches.
 d. The scale ratio tells us that every 6 mm in the scale model represents 12 m actual length, so 72 mm must represent 144 m of actual length.
 e. The scale ratio is $\dfrac{3}{5}$ or $\dfrac{1.5}{2.5}$, so a real distance of 2.5 km has a map distance of 1.5 cm.

37. This 6-year-old thinks the second cat is a smaller version of the first because he is shorter, but mathematically speaking, the cats are not similar. Notice that all of the dimensions of the first cat have not been reduced in the second cat.

SUPPLEMENTARY ACTIVITIES

1. $\frac{1}{3}$ of a pound of cheese costs \$2.50. Will $\frac{4}{3}$ as much cost more than \$2.50 or less than \$2.50?

2. There are 6 times as many dogs as cats at our SPCA. Are there more dogs or more cats?

3. Troy runs $\frac{4}{3}$ as fast as Mike. Who will win a foot race, Troy or Mike?

4. Elena's team can do a job in $\frac{2}{3}$ the time that Manny's team can do the same job. What can you say about this situation?

5. In a certain village, $\frac{2}{3}$ of the men are married to $\frac{3}{4}$ of the women. Are there more men or more women in the village?

6. Decide whether the following statements make sense or not. If the statement is correct, write "C." If a statement does not make sense, fix it by changing one of the numbers in the statement or if you cannot fix it, say why.

    a. If an orchestra can play a symphony in 1 hour, then if 2 orchestras play together, they can probably play it in $\frac{1}{2}$ hour.
    b. If it takes 3 brothers 15 minutes to drive to the soccer field to see a game, then 1 brother could drive to the game by himself in 5 minutes.
    c. If 1 boy has 3 sisters, then 2 boys probably have 6 sisters.
    d. If it takes 3 boys 2 hours to deliver papers on a certain route, then 6 boys could probably do the route in 1 hour.
    e. If a girl can walk to the mall in 20 minutes, then when she walks with 2 of her friends, they can probably get there in 60 minutes.

7. Discuss each of the following quantities listed by some sixth graders who were discussing the tractor problem. Be sure to answer these questions: Does the quantity change? Is it related to any of the other quantities? If so, how?

    a. distance between the wheels
    b. distance each wheel travels
    c. how fast the wheels turn
    d. time it takes each wheel to complete one turn
    e. the circumferences of the wheels
    f. how many times each wheel turns
    g. number of turns of the smaller wheel as compared to number of turns of the larger wheel

8. Complete each analogy and state the relationship that led you used to your answer.

   a. fire : ashes :: special occasion : ?
   b. soccer : sport :: hammer : ?
   c. sun : day :: rise : ?
   d. train : freight :: ship : ?
   e. square : octagon :: triangle : ?
   f. graceful : clumsy :: late : ?
   g. presto : instantly :: unlike : ?
   h. string : yellow ribbon :: finger : ?
   i. thieves : den :: cards : ?
   j. body : helmet :: finger : ?

9. If a train entered this tunnel, would it be able to get out the other end?

   What quantities would you use in explaining this situation to children?

10. As a fraction model, you use a rectangle with a length of 8 inches and a width of 6 inches. State the dimensions of each of the following fractional parts. Also, indicate which pieces are similar to the unit whole.

$$\frac{1}{2}, \frac{1}{3}, \frac{1}{4}, \frac{1}{6}, \frac{1}{8}, \frac{1}{12}, \frac{1}{24}$$

11. Look at each pair of pictures and decide which pairs are scale drawings.

   a.
   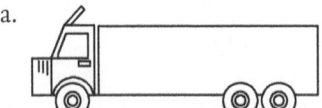   

   b.
   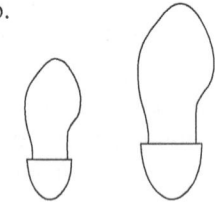

   c.

12.   a. Suppose these wheels each make 3 full turns. Will both wheels cover the
          same distance? How do you know?
      b. Suppose both wheels travel a distance of 1 mile. Will they both make the
          same number of turns? How do you know?

13.   Construct all 9 "yesterday and today" statements about each situation. Tell what
      quantity changes and how it changes.

      a. Yesterday I made juice by mixing some concentrate with some water.
      b. Yesterday I walked a certain distance in a certain amount of time.

14.   Here are two boxes, a 1″cube and a 4″cube.

      a. How many of the smaller boxes would it
          take to fill the large one?
      b. Suppose I make the edge of the smaller
          box twice its present length. How many
          small boxes would fit inside the large
          one?
      c. Suppose I made the edge of the larger box twice its present length. How
          many of the small boxes would it hold?
      d. Suppose I double the lengths of the edges of both boxes. How many small
          boxes will fit inside the large box?

15.   I have some blocks on a scale. This picture shows the scale in balance. These
      blocks may not be broken.

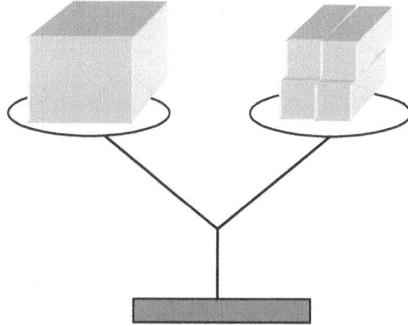

      a. What would happen if you replaced each block on the left with a block that is
          twice as heavy and you replaced each block on the right with one that is twice
          as heavy?

b. If you replaced the block on the left with one that is $\frac{3}{4}$ as heavy, what would you have to do to balance the scale?

c. If you replaced the blocks on the right with blocks that are $\frac{3}{4}$ as heavy, what would you have to do to balance the scale?

d. If you replaced the large block with one that is $\frac{2}{3}$ its present weight, what would you have to do to balance the scale?

16. In Mrs. Brown's class, there are $\frac{5}{3}$ as many students are in Mrs. Henderson's class.

a. Which teacher has more students?

b. If you know the number of students in Mrs. Henderson's class, how do you find the number of students in Mrs. Brown's class?

c. If you know the number of students in Mrs. Brown's class, how do you find the number of students in Mrs. Henderson's class?

d. If Mrs. Henderson has 27 students, how many does Mrs. Brown have?

e. If Mrs. Brown has 30 students, how many does Mrs. Henderson have?

17. In each of the following examples below, the first two strips define a relationship. Tell whether or not the second pair shows the same relationship. Explain your thinking.

a. magenta : dark green   red : magenta
b. yellow : white   orange : lime green
c. white : lime green   lime green : blue
d. dark green : blue   lime green : magenta
e. brown : orange   red : magenta
f. red : yellow   magenta : blue
g. white : lime green   lime green : dark green
h. yellow : white   orange : red

18. Are these figures similar? Explain.

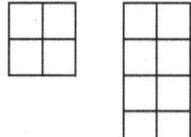

19. The edge of this cube has been multiplied by 2.

a. What happens to the area of the cube?

b. What happens to the surface area of 1 face of the cube?

20. After a brief lesson on scale, beginning sixth graders were asked to draw a knife, fork, and spoon next to a dinner plate that was approximately 5 inches in diameter. They were instructed to draw the utensils as they would look on the table, taking into account appropriate size and position. Analyze the students' drawings. Be sure to take into account all of the important proportional relationships:

a. Are the internal proportions of the utensils correct? (Here we mean the cutting part of the knife in relation to the handle, the length of the tines of the fork as compared to the handle, and the size of the cup of the spoon as compared to the handle.)

b. Are the utensils in proportion to each other?

c. Are the utensils in proportion relative to the size of the dinner plate?

21. What happens to the area of a circle if you double its radius? Triple it?

22. The perimeter of this figure is 14 units. If I scale the figure by a factor of 3, what happens to the perimeter?

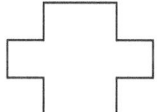

23. If two rectangles are congruent (that is, they have the same size and shape), what would be the relationship between their widths?

24. The following rectangles are similar. Find the missing measurement on the smaller rectangle.

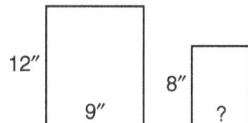

12″   9″   8″   ?

25. These letters are the same shape, but a different size. How long is the curve marked x?

26. A package of paper weighs 3 pounds. What will be the weight of a package containing the same number of sheets and the same quality of paper with sheets that are twice as long and twice as wide as the sheets in the first package?

27. a. Triangle B is similar to triangle A and A : B = 2 : 3. Draw B and label its measurements.

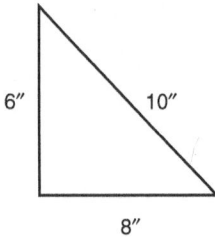

     b. Find the ratio perimeter A : perimeter B
     c. Find the ratio area A : area B.

28. A stop sign that is 6 feet high casts a shadow that is 8 feet long. The pole of a nearby streetlight casts a shadow that is 36 feet long. How high is the pole?

29. ΔABC is similar to ΔDEF. Find the lengths of the sides marked *x* and *y*.

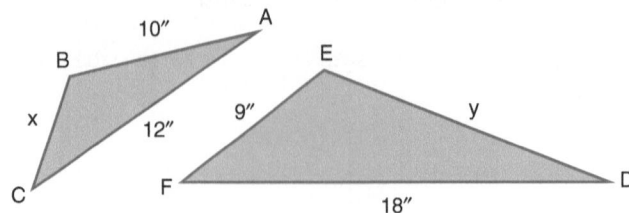

30. Find the distance across the pond. *Hint*: The similar triangles are overlapping! It might help you to draw them separately.

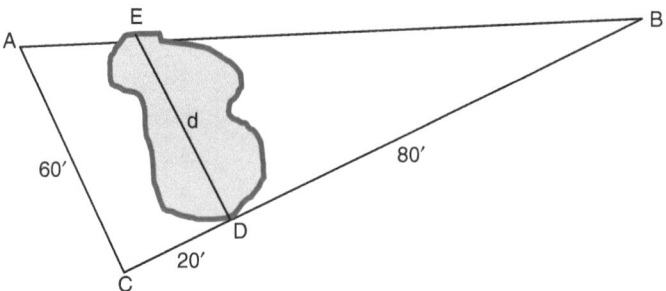

PRAXIS QUESTIONS

1. Which of the following would be an appropriate scale for the map of a city?

   a. 1mm: 100 m   b. 1 cm: 10,000 mi   c. 1 inch: 10,000,000 mi
   d. 1 inch: 3 mi   e. 1 dm: 100 mi

2. Object A is 3.5 times as tall as object B. When drawn on a scale of 3.5 feet: 5 cm, the ratio of the height of A to the height of B will be:

   a. 1 to 3.5   b. 3.5 to 1   c. 1 to 0.7   d. 0.7 to 1   e. none of these

3. If a 10-foot wide object is drawn on a scale of 2.5 feet to 2.5 cm, how wide will it be in the scale drawing?

   a. 2.5 cm   b. 2.5 ft   c. 10 cm   d. 4 cm   e. none of these

4. A 2-inch square is enlarged to become a 6-inch square. By what factor does its area increase?

   a. 3   b. 6   c. 9   d. 24   e. 36

5. An 8-foot by 10-foot rectangle is drawn using a scale of 4 feet: 2 cm. Its new measurements will be

   a. 2 feet by 5 feet   b. 4 cm by 5 cm   c. 2 feet by 2.5 feet
   d. 2 cm by 5 cm   e. none of these

6. If an 8-foot by 10-foot rectangle is scaled down in size, the 10 foot side becomes 8 inches wide. The 8-foot side will become how many inches long?

   a. 4   b. 6   c. 8   d. 10   e. none of these

7. A 10′ pole and a nearby tree both cast a shadow. About how tall is the tree?

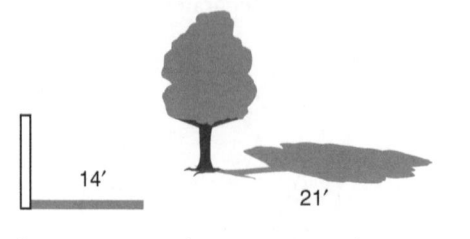

14′

21′

a. 25′   b. 15′   c. 17′   d. 20′   e. 30′

8. If school A has a student teacher ratio of 12 : 1, and school B has a smaller student–teacher ratio, B's ratio might be

a. 14 : 1   b. 14 : 2   c. 16 : 1   d. 24 : 2   e. none of these

9. A deer is 5 feet tall; a man is 6.25 feet tall; a bear is 9 feet high; a dog is 1.5 feet high; a giraffe is 16 feet tall. All were to be re-drawn on the same scale, but a mistake was made. Which one is out of proportion?

a. The dog is 3 cm tall.   b. The bear is 54 cm tall.
c. The man is 37.5 cm high.   d. The deer is 30 cm high.
e. The giraffe is 96 cm tall.

10. I drew a rectangle that measured 1.4 inches (h) by 4.2 inches (w). I locked the aspect ratio on my computer and enlarged the rectangle so that its height was 2 inches. What happened to the width?

a. stayed the same   b. changed to 4.8 inches
c. changed to 6 inches   d. changed to 5.4 inches
e. none of the above

11. I paid $16.25 admission for 1 adult and 3 children. Another person paid $26 for 2 adults and 4 children. What is the cost of an adult admission?

a. $4.50   b. $5   c. $5.50   d. $6   e. $6.50

12. Which of the following figures are not always similar to each other?

a. squares   b. circles   c. equilateral triangles
d. ellipses   e. regular hexagons

13. Which of the following rectangles is similar to a 10 × 15 rectangle?

a. 5 × 3   b. 2 × 5   c. 10 × 5   d. 6 × 10   e. 4 × 6

14. Which of the following rectangles is similar to this rectangle:

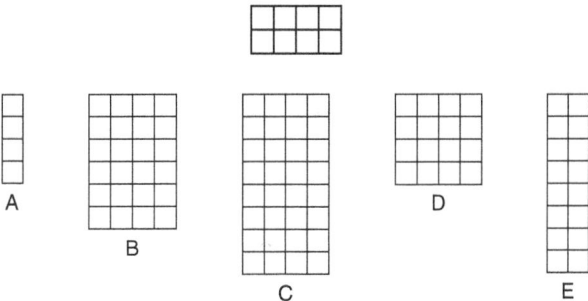

15. If two similar triangles have a scale ratio of $3:4$, what is the ratio of their perimeters?

a. $4:3$ b. $3:4$ c. $1.75:1$ d. $1:1.75$ e. none of these

## CHAPTER 5

Proportional Reasoning

DISCUSSION OF ACTIVITIES

1. All of these may constitute the unit in a fraction problem, depending, of course, on the context of the problem. The unit must be explicitly or implicitly designated or else you have ambiguity. Without context, you would not know, for example, whether part e shows a 3-unit (a unit consisting of 3 pieces) or 3 units.

2. a. 12 stars $= \frac{4}{3}$; 3 stars $= \frac{1}{3}$; 6 stars $= \frac{2}{3}$

   b. 6 balls $= \frac{3}{5}$; 2 balls $= \frac{1}{5}$; 10 balls $= \frac{5}{5} = 1$; 5 balls $= \frac{1}{2}$

   c. 12 sticks $= \frac{4}{3}$; 3 sticks $= \frac{1}{3}$; 9 sticks $= \frac{3}{3} = 1$; 5 sticks $= \frac{5}{9}$

   d. 8 blocks $= \frac{8}{3}$; 1 block $= \frac{1}{3}$; 3 blocks $= \frac{3}{3} = 1$; 6 blocks $= 2$

   e. 15 dots $= \frac{5}{6}$; 3 dots $= \frac{1}{6}$; 18 dots $= \frac{6}{6} = 1$; 9 dots $= \frac{1}{2}$; 27 dots $= \frac{1}{2}$

   f. 10 diamonds $= \frac{5}{4}$; 2 diamonds $= \frac{1}{4}$; 8 diamonds $= \frac{4}{4} = 1$; 3 diamonds $= \frac{3}{8}$

3. Both students are correct, and although they did not write down their reasoning process in the way we have done it in this chapter, each of them used a version of reasoning up and down. P went from $\frac{3}{8}$ to $\frac{1}{8}$, splitting up the 15 cards equally among the boxes in his picture. Then he extended the 5 cards per $\frac{1}{8}$ deck to all of the other boxes in order to get the number of cards in $\frac{8}{8}$. Rob used the same process, except that he went from $\frac{1}{8}$ to $\frac{2}{8}$ because he realized that he already knew how many cards were in $\frac{6}{8}$ of the deck and he only needed the number of cards in $\frac{2}{8}$ deck to complete the whole.

4. The fractions quoted by Frank and Dave refer to different units. Frank's statement used Dave's 15 pieces of pizza as the unit, while Dave used Frank's 12 pieces of pizza as the unit.

5.   11 small rectangles = $\frac{11}{3}$; 1 small rectangle = $\frac{1}{3}$; 3 small rectangles = $\frac{3}{3}$; 4 small rectangles = $\frac{4}{3} = 1\frac{1}{3}$.

6.   The same unit should be used to represent each fraction in a side-by-side representation so that the sizes of the fractions relative to each other are accurately portrayed. Because the unit is different for each fraction, a child could draw some inappropriate conclusions (e.g., $\frac{1}{2}$ and $\frac{1}{5}$ are equivalent because the same number of equal parts are shaded for both and that $\frac{2}{3}$ and $\frac{2}{4}$ are equivalent because the same number of parts are shaded for both). In order to represent all of the fractions using the same unit, that unit would have to consist of 60 blocks. Color 30, 40, 30, and 12, for $\frac{1}{2}, \frac{2}{3}, \frac{2}{4}$, and $\frac{1}{5}$, respectively.

7.   a.   Partition the two circles in the unit into 3 equal pieces each. Then you can see that the shaded amount corresponds to $\frac{1}{6}$ of the unit.

     b.   Partition the given circle into sixths; then the shaded amount is $\frac{2}{3}$ of the unit.

     c.   If you partition each piece of the unit into 3 pieces, then you can see that the shaded part is $\frac{1}{12}$ of the unit.

8.   a.

     b.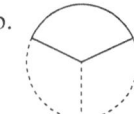

     c.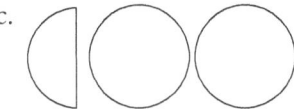

     d.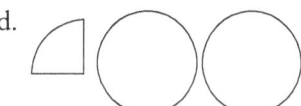

9.   The critical idea in this problem is the unit. Each person received the same amount of cake, a portion that was equivalent to $\frac{1}{6}$ of the original cake. However, as each new person takes a piece, their fraction of the cake is based on a different unit, namely, the amount of cake that remains in the pan.

10. a.   red   b.   black   c.   brown   d.   orange + red   e.   lime green

     f.   black   g.   brown   h.   magenta   i.   dark green   j.   yellow

11. a. yellow b. brown + white c. magenta d. blue
    e. lime green f. black g. yellow h. black

12. a. 2 spaces $= \dfrac{1}{4}$; 8 spaces $= 1$; 7 spaces $= \dfrac{7}{8}$

    b. 4 spaces $= \dfrac{1}{3}$; 12 spaces $= 1$; partition each of the 12 spaces into 2 equal parts, then 24 spaces $= 1$; 11 of the new spaces $= \dfrac{11}{24}$

    c. 6 spaces $= \dfrac{1}{5}$; 30 spaces $= 1$; 10 spaces $= \dfrac{1}{3}$

    d. 9 spaces $= \dfrac{3}{4}$; 3 spaces $= \dfrac{1}{4}$; 12 spaces $= 1$; 2 spaces $= \dfrac{1}{6}$; 10 spaces $= \dfrac{5}{6}$

13. The three slices of turkey Ruth bought were $\dfrac{1}{3}$ of a pound. In order to determine how much $\dfrac{1}{4}$ pound is, you must first know what the unit is. If 3 slices $= \dfrac{1}{3}$ pound, then 9 slices $= 1$ pound. If 1 pound or $\dfrac{4}{4} = 9$ slices, then $\dfrac{1}{4}$ pound must be $2\dfrac{1}{4}$ slices.

14. a. yellow $= 3$; b. blue $= \dfrac{2}{9}$; c. $1 = 2$ yellows;

    d. yellow; e. green $= 1\dfrac{1}{2}$, red $= 4\dfrac{1}{2}$, and yellow $= 9$.

15. The first question we need to ask as we consider this problem is "What is the unit?" We should notice that a small pizza and a medium pizza were ordered, so that we cannot combine them and say there were 16 slices. Those 16 slices are different sizes! The unit is 8 small slices + 8 medium slices. How much was eaten? The only way to say it is $\dfrac{2}{8}$ or $\dfrac{1}{4}$ of the small pizza and $\dfrac{3}{8}$ of the medium pizza.

16. 16 liters $= \dfrac{2}{5}$; 8 liters $= \dfrac{1}{5}$; 40 liters $= 1$ tank (full).

17. Every one of the whole pies is $3\left(\dfrac{1}{3}\text{-pies}\right)$, so altogether the manager has 22 $\left(\dfrac{1}{3}\text{-pies}\right)$ or $11\left(\dfrac{2}{3}\text{-pies}\right)$.

18.   a.    b.    c.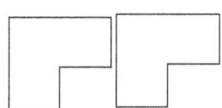

d. 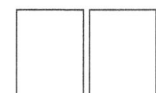   e.

19.   a. $5\frac{4}{6}$(6-packs)   b. $1\frac{10}{24}$(24-packs)   c. $1\frac{16}{18}$(18-packs)

d. $4\frac{2}{8}$(8-packs)   e. $68\left(\frac{1}{2}\text{-cans}\right)$ f. $136\left(\frac{1}{4}\text{-cans}\right)$

g.   17(pairs)

20.   a.   Look at the rows in the picture.   b.   Look at the columns in the picture.
c.   Look at individual hearts.   d.   Look at pairs of hearts.
e.   Look at half hearts.

21.   Some possibilities are:

a. $15\,\text{stars} = 3(5\text{-packs}) = 7\frac{1}{2}(\text{pair}) = 1\frac{3}{12}(\text{dozen}) = 30\left(\frac{1}{2}\text{-stars}\right)$

b. $16\,\text{colas} = 8(\text{pair}) = 48\left(\frac{1}{3}\text{-cans}\right) = \frac{1}{2}(32\text{-pack})$

c. $26\,\text{eggs} = 2\frac{2}{12}(\text{dozen}) = 52\left(\frac{1}{2}\text{-eggs}\right) = 8\frac{2}{3}(3\text{-packs})$

22.   a.   Look at the medium rectangle.
b.   Look at the 9 small squares or the large square or the 4 medium squares.
c.   Look at a set of 3 small squares.
d.   Look at one of the small rectangles.
e.   Look at a pair of 2 medium squares.

23.   a. $\frac{1}{4}$ b. $\frac{1}{18}$ c. $\frac{1}{3}$ d. $\frac{1}{8}$ e. $\frac{1}{54}$ f. $\frac{1}{24}$

g. $\frac{1}{16}$ h. $\frac{1}{6}$

24.   a.   If three pints cost $1.59, then 1 pint should cost $0.53 $\left(\frac{1}{3}\text{ as much}\right)$ and 4

should cost $2.12 (4 times the cost of 1). The milk cartons are priced

proportionally. $\dfrac{2.12}{4} = \dfrac{0.53}{1}$ and $\dfrac{1.59}{3} = \dfrac{0.53}{1}$

b. This situation does not involve proportional or inversely proportional relationships. How long it takes to drive to the basketball game has nothing to do with how many people are in the car.

c. The number of people is cut in half, which means that the work should take twice as long. This is the case, and the situation involves an inversely proportional relationship between the number of people and the time it takes to do a job.

d. This situation does not involve any proportional or inversely proportional relationships. The number of sisters one boy has has nothing to do with how many sisters another boy has.

e. This situation does not involve any proportional or inversely proportional relationships. Time spent on solving a problem is not really dependent on where a person is solving it.

f. If the number of eggs and the time it takes to eat them are proportional, then it should take 20 times as long to eat 20 eggs as it does to eat one egg. However, this is not the case. 1 egg per 20 seconds = 3 eggs per minute. The average rate for eating 20 eggs was 4 eggs per minute.

g. Gallons of gas used and distance traveled are proportionally related. A full tank of gas (15 gallons) is $3\frac{1}{3}$ times $4\frac{1}{2}$ gallons and 333 miles is $3\frac{1}{3}$ times 100 miles, the distance traveled on $4\frac{1}{2}$ gallons.

h. Amount spent and sales tax are proportionally related. When you spent $35 and paid $2.10 in sales tax, you spent 7 times as much as when you spent $5 and paid 7 times as much as you paid on sales tax.

25. a.

	# tickets	Profit ($)	Notes
a	1	1.15	given
b	100	115	100a
c	10	11.50	b÷10
d	5	5.75	$\frac{1}{2}c$
e	20	23	2c
f	3	3.45	3a
g	128	147.20	b + d + e + f

The school makes $147.20 on 128 tickets.

b.

	# cans	Cost ($)	Notes
a	1	4.49	given
b	10	44.90	10a
c	5	22.45	$\frac{1}{2}b$
d	15	67.35	b + c

There is enough money to buy 15 cans.

c.

	Distance (km)	Time (min)	Notes
a	10	45	given
b	1	4.5	a ÷ 10
c	6	27	6b
d	0.1	0.45	b ÷ 10
e	0.2	0.90	2d
f	0.01	0.045	d ÷ 10
g	0.05	0.225	5f
h	6.25	28.125	c + e + g

It would take him just a little over 28 minutes.

26. a.

Words	Minutes
575	15
2300	60
2875	75

b.

Boxes	Dollars
3	6.88
15	34.40
1	2.2933
2	4.59
17	38.99

c.

Caps	Points
1	60
10	600
9	540
8	480

d.

Inches	Miles
1	195
2	390
0.1	19.5
0.01	1.95
0.02	3.9
0.001	0.195
0.005	0.975
2.125	414.375

e.

Mac	Dad	Total
$3	$5	$8
$30	$50	$80
$15	$25	$40
$45	$75	$120

f.

Girls	Quarts
5	3.5
10	7
1	0.7
4	2.8
14	9.8

27. a. Distance and area are not proportional.

Diameter (feet)	Area (sq ft)
3	7.06858
6	28.27433

×2   ×4

b. Distance and cost are not proportional.

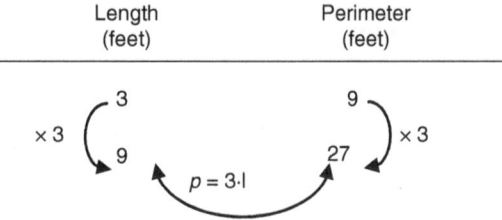

Distance (miles)	Cost (dollars)

×5 (2 → 10)   (5.50 → 21.50) × 3.91

×5 (50)   (101.50) × 4.72

c. There is a proportional relationship between the length and perimeter.

Length (feet)	Perimeter (feet)

×3 (3 → 9) p = 3·1 (9 → 27) × 3

28. Hold the number of cords constant and find out how long it would take 4 men to cut them. Then hold the number of men constant and find the amount of time it would take them to cut 3 cords.

Men	8	4	4
Cords	9	9	3
Hours	6.5	13	$\frac{13}{3} = 4\frac{1}{3}$

29. Hold the time constant and figure out how many cars 14 robots can make. Then hold the robots constant and figure out how many cars they can make in 8 hours.

Robots	3	12	2	14	14
Cars	19	76	$\frac{76}{6}$	$76 + \frac{76}{6}$	$\frac{76}{6} + \frac{76}{30} = 17$
Hours	40	40	40	40	8

Note that the answer is greater than 17, but less than 18. Therefore we conclude that they can make 17 cars.

30. a. There is a proportional relationship between weight in pounds and weight in kilograms. In both of the instances given, the ratio of pounds to kilograms is 2.20. We can express this in the equation $p = 2.2\,k$.

    b. There is a proportional relationship between the striking distance and the time until you hear the crash. The time in seconds at which the crash is heard is three times the distance in miles from where the lightning struck. We can express this in the equation $t = 3d$.

    c. There is a proportional relationship between the pressure on your ears and your distance under water. The pressure is 0.43 times the distance. We can express this in the equation $p = 0.43d$.

31. a.

Weeks	Horses	Pounds
3	4	45
1	4	15
4	4	60
4	1	15

    b.

Robots	Parts	Minutes
5	5	5
10	10	5
10	100	60 (1hr)
10	1000	600 (10 hr)

    c.

#hens	#eggs	#days
$1\frac{1}{2}$	$1\frac{1}{2}$	$1\frac{1}{2}$
3	3	$1\frac{1}{2}$
3	6	3
24	48	3
24	384 (or 32 doz)	24

32. a. $\frac{1}{5}$ of 50% = 10%   b. $\frac{1}{4}$ of 25% = $6\frac{1}{4}$%   c. $\frac{1}{8}$ of 50% = $6\frac{1}{4}$%

    d. $\frac{3}{4}$ of 20% = 15%

33. a. 100% of 80 = 80, so 25% of 80 = 20

    b. Start with $\frac{15}{50}$ and scale up to $\frac{x}{100}$, then $\frac{15}{50} = \frac{30}{100} = 30\%$

    c. Start with $\frac{45}{60}$ and scale up to $\frac{x}{100}$.   d. 20 per 100 = ? per 70

#	Out of
45	60
15	20
30	40
75	100

#	Out of
20	100
10	50
2	10
4	20
14	70

e. Begin with 65 out of 100.

#	Out of
65	100
13	20
52	80
6.5	10
58.5	90

f. Begin with 40 out of 100 and work until you get 28 out of something.

x	Out of
40	100
20	50
8	20
28	70

g. 15 out of 100 = ? out of 80

#	Out of
15	100
3	20
12	80

h. 36 out of 90 = ? out of 100

x	Out of
36	90
4	10
40	100

i. Begin with 45 out of 100 and find 27 out of ?

#	Out of
45	100
9	20
27	60

34. a. $1\% = \dfrac{1}{100}; 2\% = \dfrac{2}{100} = \dfrac{1}{50}; 6\% = \dfrac{3}{50}$

b. $50\% = \dfrac{1}{2}$, so $\dfrac{3}{2}$ must be 150%

c. $1\% = \dfrac{1}{100}; 2\% = \dfrac{2}{100} = \dfrac{1}{50};$ so $\dfrac{20}{50}$ must be 50%. Or, you could have noticed that $\dfrac{25}{50}$ reduces to $\dfrac{1}{2}$, which is 50%.

d. $\dfrac{1}{5} = 20\%$, so $\dfrac{3}{5} = 60\%$

e. $1 = 100\%$, so $3 = 300\%$

f. $\dfrac{1}{8} = \dfrac{1}{4}$ of $\dfrac{1}{2}$, so $\dfrac{1}{4}$ of $50\% = 12\dfrac{1}{2}\%$

g. $\dfrac{1}{6} = \dfrac{1}{3}$ of $\dfrac{1}{2}$, and $\dfrac{1}{3}$ of $50\% = 16\dfrac{2}{3}\%$

h. Using the result of the last problem (g), $\dfrac{5}{6} = 5\left(16\dfrac{2}{3}\%\right)$. $5(16) + 5\left(\dfrac{2}{3}\right) = 80 + 3\dfrac{1}{3} = 83\dfrac{1}{3}\%$.

i. $\dfrac{1}{12} = \dfrac{1}{3}$ of $\dfrac{1}{4} = \dfrac{1}{3}$ of $25\% = 8\dfrac{1}{3}\%$. so $\dfrac{7}{12} = 7\left(8\dfrac{1}{3}\%\right) = 56 + \dfrac{7}{3} = 58\dfrac{1}{3}\%$

35. Remember that percents indicate a proportion (a relative amount), not an absolute number. The hospitals could have had different numbers of babies born that day, yet both could have had 50% male newborns. For example, 50% of 20 = 10 and 50% of 14 = 7. Jack could be correct if both hospitals had the same number of births that day.

36. a. 80 out of 100 is like 40 out of 50.
    b. 1% would be $150, so 6% must be $900.
    c.

#	$
1	18
$\frac{1}{2}$	9
4	72
4.5	81

    d. 1% (200) = 2 lb, so 85% would be 85(2) or 170 pounds of water.

37. a. 10% is $1.68; 5% is $0.84; 15% is $2.52
    b. $4.64 + $2.32 = $6.96
    c. $12.56 + $6.28 = $18.84
    d. $6.22 + $3.11 = $9.33

38. I am going to set up a ratio table that contains all of the components I will need to determine the percentages.

%	$
1	0.125
0.1	0.0125
0.4	0.05
0.5	0.0625
5	0.625
7	0.875
10	1.25
20	2.50
40	5.00

    Using the table, we can get the percentages we need, rounding to the nearest cent.
    a. 80% = $10.00
    b. 47% = $5.88
    c. 200% = $12.50 + $12.50 = $25
    d. 120% = $12.50 + $2.50 = $15
    e. 1.5% = $0.19
    f. 60% = $7.50
    g. 5.4% = $0.68
    h. 350% = $43.75

SUPPLEMENTARY ACTIVITIES

1. Solve by reasoning up and down.

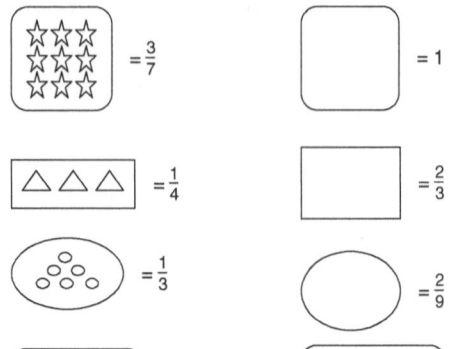

2. Tell what fraction each x represents. Solve by reasoning up and down.

a.

0   x $\frac{2}{3}$

b.

0    $\frac{1}{5}$ x

c.

0   x $\frac{9}{4}$

d.

0        $1\frac{1}{2}$ x

3. Find the missing component of each statement.

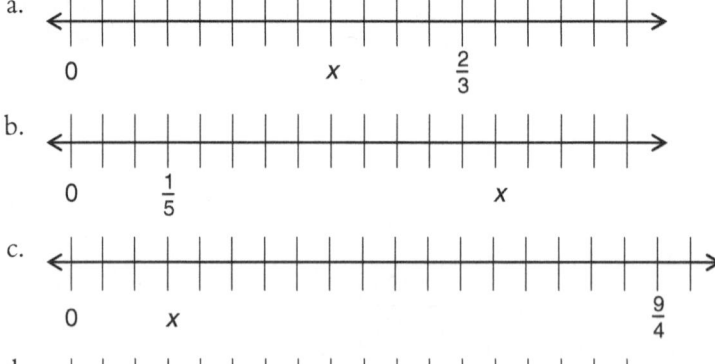

a. ⊢————⊣ is $\frac{5}{3}$ of

b. _____ is $\frac{1}{4}$ of ▢

c. 3 pounds is _____ of 4 ounces

d. 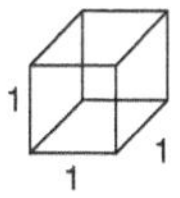 is _____ of

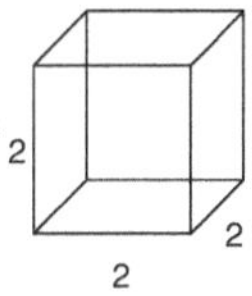

e.  is $1\frac{2}{3}$ of _____

f. 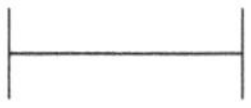 is $\frac{2}{3}$ of

g.  is _____ of

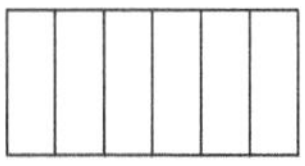

h. _____ is $\frac{2}{3}$ of

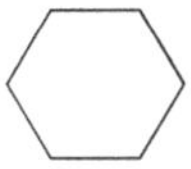

i. 350 cm is _____ of 42 dm

j. is $\frac{5}{4}$ of

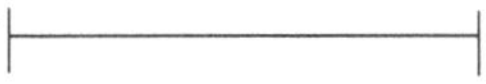

k. 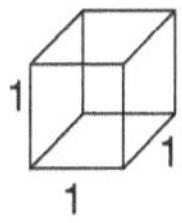 is $\frac{1}{12}$ of

l. is $\frac{1}{4}$ of

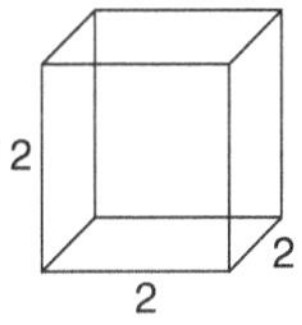

m.  is _____ of 25 g

n. ☆☆ is $\frac{2}{7}$ of _____

o. 2 carats is _____ of $\frac{1}{3}$ carat

4.   Use pattern pieces to complete this chart.

Green	Blue	Red	Yellow
1			
			3
		2	
			$\frac{2}{3}$
			$1\frac{1}{2}$
$\frac{1}{10}$			
		$\frac{3}{8}$	
		4	
	$\frac{1}{9}$		
		$\frac{2}{3}$	

6.   Trace your pattern pieces to answer the following questions.

a.   △ $= \frac{1}{8}$. Draw $\frac{3}{4}$.

b.   ⬡◁ $= 2$. Draw $\frac{2}{3}$.

c.   ⬡⬡ $= 1$. What does ▱△ represent?

d.   ▱▱ $= \frac{1}{4}$. What does ⬡⬡ represent?

e.   ▱△ $= \frac{1}{5}$. Draw $\frac{1}{3}$.

f.   ▱△ $= \frac{1}{3}$. Draw $\frac{5}{6}$.

7.   ▱ $= \frac{1}{6}$ Perform this operation using the pattern pieces. $\frac{1}{6} \div \frac{1}{3} = ?$ (*Hint.*

First find 1, then determine $\frac{1}{3}$. How many of the $\frac{1}{3}$ pieces are in $\frac{1}{6}$?)

8. $\boxed{\frac{1}{4}} + \frac{1}{6} = ?$ $\left(\textit{Hint}:\text{ Find 1, then determine }\frac{1}{6}.\text{ Put the }\frac{1}{4}\text{ and the }\frac{1}{6}\text{ pieces}\right.$

    on the unit to determine the sum.$\Big)$

9. = 1. Perform this division using the pattern pieces: $\frac{3}{4} \div \frac{1}{6}.$ $\left(\text{How}\right.$

    many $\frac{1}{6}$-pieces are in $\frac{3}{4}$?$\Big)$

10. = 1. Divide: $1\frac{5}{6} \div \frac{1}{3}$

11. $= \frac{1}{2}.$ Use the pattern pieces to add: $\frac{1}{4} + \frac{1}{12}.$

12. $= \frac{1}{2}.$ Perform this addition: $\frac{5}{12} + \frac{11}{12}.$

13. $= \frac{1}{4}.$ Perform this subtraction: $2 - \frac{5}{6}.$

14. What operations with the pattern pieces are analogous to finding a common denominator for adding fractions?

15. How do you use pattern pieces to reduce a fraction to lowest terms?

16. Terry solved the following problem. Do you think he was correct? Ask yourself: What is the unit?

    Here is a day's supply of orange juice. You have taken only one drink of juice. How much of your day's supply do you still have left?

Terry

I measure it out and get $\frac{1}{5}$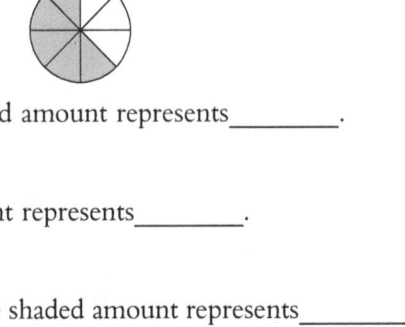
but it is $\frac{1}{5}$ of $\frac{1}{2}$ of all my juice that
I drank.
$\frac{1}{5}$ of $\frac{1}{2} = \frac{1}{10}$ drank
$\frac{9}{10}$ left

17.   Name the shaded portion of this picture in relation to each unit.

a. If the unit is ⬭⬭ the shaded amount represents_____.

b. If the unit is ◠ the shaded amount represents_____.

c. If the unit is ◯◯◯ the shaded amount represents_____.

d. If the unit is ◠◯ the shaded amount represents_____.

18.   I have $\frac{5}{6}$ of a pie in my refrigerator. If I eat half a pie, how much will I have left?
Analyze the work of the two students shown here.

Andy    I cut each piece in half
Here is what I eat

There is $\frac{5}{12}$ left.

Barb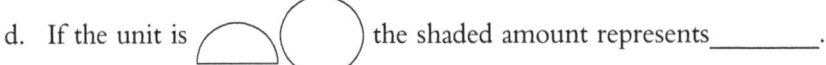
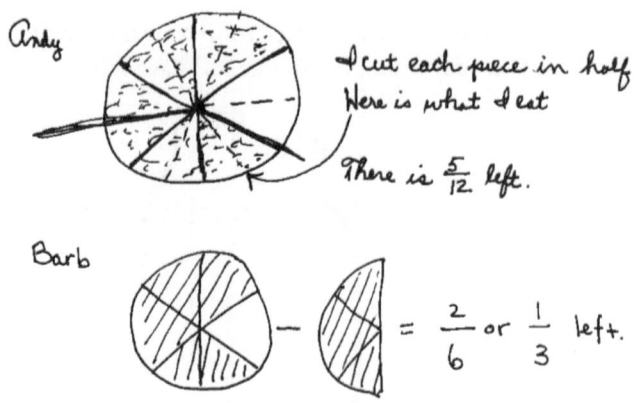 $= \frac{2}{6}$ or $\frac{1}{3}$ left.

19. Tim spent $\frac{1}{4}$ of his money, and Pete spent half of his. How could it be possible that Tim spent more money than John spent?

20. Comment on this picture:

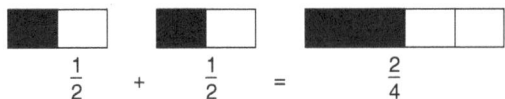

21. Four friends wanted a snack. Mrs. Johnson had 3 large cookies. Amy planned to split them this way:

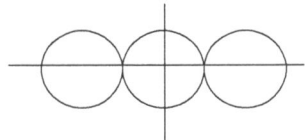

Seth planned to split them this way:

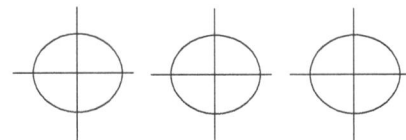

a. Name the share that each gets using Amy's plan.
b. Name the share that each person gets using Seth's plan.

22. Tell the size of chunk that helps you to "see" each fraction below.

a. $\frac{1}{4}$ b. $\frac{1}{10}$ c. $\frac{1}{20}$ d. $\frac{1}{5}$

23. Unitize each quantity in at least two different ways:

a. 3 weeks =
b. \$0.05 =
c. 16 (half pints) =
d. $1\frac{1}{2}$ dozen =
e. 26 (quarter-miles) =
f. 3 inches =
g. 5 (5-packs) of gum =

> h. 17 pairs of shoelaces =
> i. 3(8-packs) of soda =

24. Find the unit rate.

    a. Three apples cost $1.97.   b. Four movie tickets cost $39.80.
    c. Two shirts cost $29.90.   d. 24 cans of soda cost $11.98.
    e. 6 sticks of gum cost $0.69.

25. Find the better deal (considering price only) without using a unit rate.

    a. $35 to go 80 miles by train or $45 to go 120 miles by bus.
    b. work 8 hours for $48 or work 12 hours for $60.
    c. 6 car washes for $32 or 18 car washes for $94.
    d. $5000 in taxes on a $200,000 home or $4000 in taxes on a $150,000 home.
    e. a 6-ounce package for $0.54 or an 18-ounce package for $1.88.
    f. 16 issues of POP magazine for $10 or 24 issues of Max magazine for $18.
    g. $58 for newspaper delivery for 28 weeks or $70 for 35 weeks of delivery.
    h. 18 family-size cans of soup for $42 or 24 of the same size for $54.

26. Without doing any pencil-and-paper calculating, tell which is the better deal (considering only price).

    a. 2 gallons of milk for $4.45 or 6 gallons of milk for $13.50.
    b. $12 tickets for $52 or 3 for $12.50.
    c. 3 apples for $2.50 or 3 apples for $2.10.
    d. 16 photocopies for $1.44 or 36 photocopies for $3.06.
    e. 27 loads of wash for $60 or 36 loads for $72.
    f. $2.40 for 6 lb of squash or 1.5 lb of squash for $1.23.
    g. 1600 bushels of soybeans per 9 acres or 2050 bushels of soybeans per 15 acres.
    h. 20 boxes of candy for $18 or 32 boxes of the same candy for $24.80.

27. Do this in your head: Which is the better buy: a 6-pack of cola for $1.89 or a 12-pack for $3.29?

28. Do this in your head: Notebook paper costs $1.89 for 500 sheets or $0.99 for 250 sheets. Which is the better buy?

29. Do this in your head: A 10-pound bag of rice costs $5.50. A 15-pound bag costs $8.40. Which is the better buy?

30. Solve by reasoning up and down.

    a. In 3 weeks, a horse eats 10 pounds of hay. How much will he eat in 5 weeks?

b. A dragon's tail is $\frac{3}{4}$ of its total length. If the tail is $6\frac{3}{4}$ feet long, what is the total length of the dragon?

c. Mary lost $\frac{3}{7}$ of her baseball cards. She has 16 left. How many did she lose?

d. It takes Jim and his 2 brothers 4 hours to do the yard work. When one of the boys is sick, how long does it take the other two to do the work?

e. You watch TV for $12\frac{1}{2}$ hours a week. How many hours do you watch in a year?

f. It takes Steve $1\frac{1}{3}$ hours to mow his lawn. He spent 100 hours mowing last summer. How many times did he mow?

g. A 12-person cleaning crew cleans a certain office building every night. It takes them 6 hours to complete the job. One night 2 people failed to show up for work. How long did it take the rest of the crew to finish the job?

h. Mark studies $2\frac{3}{4}$ hours a week for each credit that he carries during a semester. This semester, he is taking 15 credits. How much time does he spend studying?

i. The home improvement store has fencing for $16 for 2.5 lineal feet. You figure that it will take 110 feet of fencing to fence your garden. How much will it cost you?

j. Jack drove 150 miles on 5 gallons of gas. How many gallons would his car use if he drove 225 miles?

k. If it takes 15 hours to fill a swimming pool with one hose, how long will it take if I use 4 hoses (same hoses, water running at the same rate through each)?

l. If it takes 3 painters 5 days to paint 3 houses, how long will it take 9 painters to paint them?

m. At Camp Getaway, 5 boys use 3 bottles of shampoo in 2 weeks. At the same rate, how much shampoo will 30 boys use in a week?

31. Josh and Kristin are seventh grade students who are using reasoning to solve a fairly complex problem. See if you can figure out what they are doing. Also, account for the discrepancy in their answers.

For every 50 people who attend the school fair, about 37 of them will purchase a raffle ticket. The school makes $1.25 profit on every raffle ticket. If 723 people go to the fair, how much money can the school hope to make on the raffle?

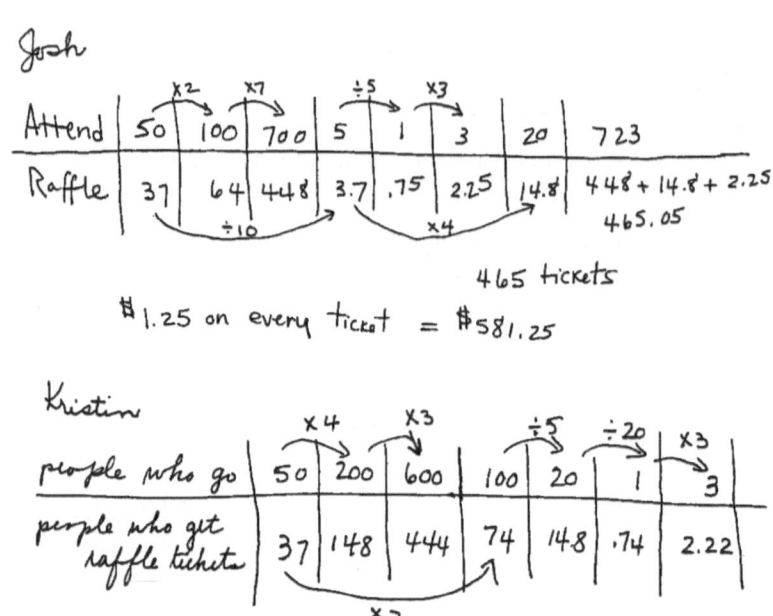

465 tickets

$1.25 on every ticket = $581.25

723 = 444 + 74 + 14.8 + 2.22 = 535.02

535 tickets $1.25 each   = $668.75

32. A dragon's tail is $\frac{5}{18}$ of its total body length. If the tail is 15 feet long, find the total length of the dragon in feet.

33. Another chicken problem. A chicken and a half lays an egg and a half in a day and a half. How many dozen eggs do 12 chickens lay in 12 days?

34. If 3 robots can assemble 17 cars in 10 minutes, how many cars can 14 robots assemble in 45 minutes?

35. Return to chapter 1 and solve the Lewis Carroll problem.

36. Use a proportion table to solve this problem:
   Chicken costs $2.69 per pound. Marie buys 3.62 pounds. How much will she pay?

37. Solve by reasoning aloud.

   a. I have a hole in my pocket and I lost $\frac{2}{3}$ of the coins I put into my pocket. I have only 14 coins left. How many coins did I start with?

b. Rhonda and Marty are sharing a pizza cut into equal-sized pieces. Rhonda ate $\frac{1}{3}$ of the pizza and Marty ate $\frac{1}{2}$ of the pizza. If 3 pieces remain, how many pieces did Rhonda eat?

38. Use benchmark percents to help you write each fraction as a percent.

a. $\frac{4}{50}$ b. $\frac{21}{25}$ c. $\frac{1}{12}$ d. $\frac{11}{12}$ e. $\frac{1}{8}$ f. $\frac{3}{8}$

g. $\frac{9}{25}$ h. $\frac{15}{50}$ i. 4   j. $\frac{5}{4}$ k. $\frac{9}{8}$

39. Use a ratio table to help you reason up/down to obtain the following percentages of $8500.

a. 4%   b. 85%   c. 40%   d. 120%   e. 2.5%   f. 28%

40. Answer these questions by reasoning.

a. 15% of 80 = _____   b. 15 = _____% of 50
c. 30% of _____ = 21   d. 20% of 70 = _____
e. 20 = 40% of _____   f. 5 = 20% of _____
g. 15 = _____% of 40   h. 30% of _____ = 24
i. 3 = 5% of _____   j. _____ is 2.5% of 80
k. _____ is 125% of 30   l. 6 = _____% of 50
m. 300 = _____% of 75

41. The University Book Station prices books according to the following equation. C represents the cost of the book to the bookstore and S represents the price at which they sell the book to students. $\frac{C}{0.75} = S$. What is the percent of markup on the books?

PRAXIS QUESTIONS

1. If ⬚⬚ $= \frac{1}{3}$, what fraction of the picture below is shaded?

a. three   b. two and one-half
c. two and three-fourths   d. eleven twelfths
e. not given

2. If 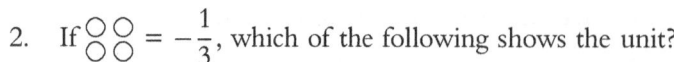 $= -\dfrac{1}{3}$, which of the following shows the unit?

    a.

    b.

    c.

    d.

    e. not given

3. A farmer has enough grain to feed 50 cattle for 10 days. If he sells 10 head of cattle, how many full days will the same amount of food last?

    a. 16   b. 13   c. 12   d. 18   e. 14

4. You have enough money to buy 30 donuts or 20 colas. If you buy only 18 donuts, how many colas can you buy?

    a. 6   b. 8   c. 10   d. 12   e. 14

5. How many marbles are in $\dfrac{7}{6}$ of this collection?

    a. 21   b. 24   c. 30   d. 36   e. none of these

6. Mr. Rich spent $\dfrac{2}{7}$ of his fortune on a new home. If \$2.7 million is $\dfrac{3}{5}$ of his fortune, the amount he spent on his new home is

    a. less than a million   b. about \$1 million
    c. about \$2 million   d. between \$1.5 and \$2 million
    e. between \$1 and \$1.5 million

7. If a 12-ounce box of pasta costs \$0.96 and a 18-ounce box costs \$1.20, which procedure would NOT be appropriate for finding the best deal?

    a. $\dfrac{0.96}{6}$ and $\dfrac{1.20}{9}$ b. $\dfrac{0.96}{12}$ and $\dfrac{1.20}{18}$ c. $\dfrac{0.96}{4}$ and $\dfrac{1.20}{4}$

    d. $\dfrac{0.96}{2}$ and $\dfrac{1.20}{3}$ e. $\dfrac{12}{0.96}$ and $\dfrac{18}{1.2}$

8. What percent of 125 is 105?

    a. 84%   b. 0.84 %   c. 119%   d. 1.19%   e. none of these

9. Which of the following fractions is equal to 0.25%?

   a. $\dfrac{1}{40}$ b. $\dfrac{1}{4}$ c. $\dfrac{5}{2}$ d. $\dfrac{1}{400}$ e. $\dfrac{50}{2}$

10. Bob earns \$250 per week. If he spends 20% of his income on rent, 25% for food, and 10% for savings, how much is left for other expenses?

    a. \$125   b. \$140   c. \$112.50   d. \$132.50
    e. \$137.50

11. $\dfrac{1}{4}$% written as a decimal is

    a. 25.0   b. 2.5   c. 0.25   d. 0.025   e. 0.0025

12. Jack pays \$520 a month for rent, and his monthly paycheck after taxes is \$1300. Which computation shows the percent of Jack's paycheck that is used to pay rent?

    a. $(1300 \div 520) \cdot 100$ b. $(520 \div 1300) \cdot 100$ c. $(5.2 \cdot 1300) \cdot 100$
    d. $(13 \cdot 520) \cdot 100$ e. $(5.2 \cdot 13) \cdot 100$

13. If x is 65% of 400, then what is the value of x?

    a. 660   b. 540   c. 260   d. 200   e. 140

14. 40% of what number is 300?

    a. 750   b. 500   c. 450   d. 600   e. 120

15. If 20% of a number is 8, what is 25% of the number?

    a. 2   b. 10   c. 12   d. 11   e. 15

## CHAPTER 6

Reasoning with Fractions

DISCUSSION OF ACTIVITIES

1. A method is suggested for each problem; however, in some cases, other methods may be used.

    a. CB $\dfrac{8}{14}$ is greater than $\dfrac{1}{2}$ and $\dfrac{4}{9}$ is less than $\dfrac{1}{2}$. So $\dfrac{8}{14} > \dfrac{4}{9}$.

    b. SNP $\dfrac{3}{17}$ and $\dfrac{3}{19}$ each have the same number of pieces, and so the critical question becomes, "What size are the pieces?" Seventeenths are larger than nineteenths, so $\dfrac{3}{17} > \dfrac{3}{19}$.

    c. SSP $\dfrac{8}{13}$. All of the pieces are the same size, so if you have 8 of them you have more than if you have 5 of them. $\dfrac{8}{13} > \dfrac{5}{13}$

    d. CB $\dfrac{3}{2}$ is $\dfrac{1}{2}$ greater than 1, but $\dfrac{4}{3}$ is only $\dfrac{1}{3}$ greater than 1. $\dfrac{3}{2} > \dfrac{4}{3}$

    e. $\dfrac{2}{3}$ SNP   f. $\dfrac{7}{9}$ CB$\left(\dfrac{1}{2}\right)$ g. $\dfrac{7}{8}$ CB(1)   h. $\dfrac{1}{5}$ SNP

    i. $\dfrac{3}{4}$ unit fractions: $\dfrac{5}{9} = \dfrac{1}{3} + \dfrac{1}{9} + \dfrac{1}{9}$ $\dfrac{3}{4} = \dfrac{1}{3} + \dfrac{1}{6} + \dfrac{1}{4}$ Both fractions contain $\dfrac{1}{3}$, but the remaining fractions composing $\dfrac{3}{4}$ are greater than the remaining fractions in $\dfrac{5}{9}$. $\dfrac{3}{4} > \dfrac{5}{9}$.

    j. $\dfrac{3}{5}$ CB$\left(\dfrac{1}{2}\right)$ k. $\dfrac{4}{9}$ unit fractions   l. $\dfrac{11}{21}$ CB$\left(\dfrac{1}{2}\right)$ m. $\dfrac{5}{8}$ CB$\left(\dfrac{1}{2}\right)$

    n. $\dfrac{5}{11}$ CB$\left(\dfrac{1}{2}\right)$ o. $\dfrac{4}{9}$ unit fractions   p. $\dfrac{7}{12}$ CB$\left(\dfrac{1}{2}\right)$ q. $\dfrac{3}{7}$ CB$\left(\dfrac{1}{2}\right)$

    r. $\dfrac{5}{9}$ CB$\left(\dfrac{1}{2}\right)$ s. $\dfrac{10}{11}$ CB(1)   t. $\dfrac{13}{14}$ CB(1)

d. First create the largest fraction, and then with the remaining numbers, the smallest fraction.

$$\frac{8}{5} - \frac{6}{7}$$

e. Choose the denominators to give the smallest-size pieces. After you choose 8 and 7 as denominators, it does not matter which numerator is 5 and which is 6.

$$\frac{5}{8} \times \frac{6}{7} \text{ or } \frac{6}{8} \times \frac{5}{7}$$

f. You want the largest possible numerators, so choose 8 and 7. Then 5 and 6 may be placed in either of the denominators.

$$\frac{8}{6} \times \frac{7}{5} \text{ or } \frac{8}{5} \times \frac{7}{6}$$

g. The divisor must be as large as possible, so choose $\frac{8}{5}$ as the divisor. Then the dividend should be as small as possible:

$$\frac{6}{7} \div \frac{8}{5}$$

h. The dividend must be as large as possible, so choose $\frac{8}{5}$. The divisor should be as small as possible.

$$\frac{8}{5} \div \frac{6}{7}$$

5. a. The fraction gets larger.   b. You cannot tell.
    c. The fraction gets smaller.   d. The fractions are equivalent.

6. There is a difference of $\frac{3}{5}$ and we need 3 fractions (4 equal spaces) between $\frac{2}{5}$ and 1. This means that the spaces between the fractions will be $\frac{3}{20}$. Obtain the fractions by adding $\frac{3}{20}$, $\frac{6}{20}$, and $\frac{9}{20}$, respectively, to $\frac{2}{5}$. The fractions are $\frac{11}{20}$, $\frac{14}{20}$, and $\frac{17}{20}$.

7. If a is replaced by a positive fraction, say $\frac{3}{4}$ for example, then $\frac{1}{a}$ will be positive.

$$\frac{1}{\frac{3}{4}} = \frac{4}{3} = 1\frac{1}{3}$$

2. a. $\dfrac{7}{8} = \dfrac{1}{3} + \dfrac{1}{6} + \dfrac{1}{4} + \dfrac{1}{8}$ $\dfrac{3}{5} = \dfrac{1}{5} + \dfrac{1}{5} + \dfrac{1}{10} + \dfrac{1}{10}$ $\dfrac{7}{8} > \dfrac{3}{5}$

   b. $\dfrac{7}{8} = \dfrac{1}{2} + \dfrac{1}{4} + \dfrac{1}{8}$ $\dfrac{5}{6} = \dfrac{1}{2} + \dfrac{1}{4} + \dfrac{1}{12}$ $\dfrac{7}{8} > \dfrac{5}{6}$

   c. $\dfrac{7}{8} = \dfrac{1}{2} + \dfrac{1}{3} + \dfrac{1}{24}$ $\dfrac{9}{10} = \dfrac{1}{2} + \dfrac{1}{3} + \dfrac{1}{15}$ $\dfrac{9}{10} > \dfrac{7}{8}$

3. a. Write $\dfrac{1}{6}$ as $1 \div 6$ and $\dfrac{1}{5}$ as $\dfrac{6}{5} \div 6$. Then any fractions between 1 and $1\dfrac{1}{5}$ (divided

   by 6) will be between $\dfrac{1}{6}$ and $\dfrac{1}{5}$. e.g., $\dfrac{1\frac{1}{6}}{6} = \dfrac{7}{36}, \dfrac{1\frac{1}{7}}{6} = \dfrac{8}{42}$, and $\dfrac{1\frac{1}{8}}{6} = \dfrac{9}{48}$.

   b. $\dfrac{1}{13} = \dfrac{14}{13} \div 14 = 1\dfrac{1}{13} \div 14$, so $\dfrac{7}{13} = \dfrac{7\frac{7}{13}}{14}$. To find fractions between $\dfrac{7}{14}$ and

   $\dfrac{7\frac{7}{13}}{14}$ choose numerators greater than 7 and less than $7\dfrac{7}{13}$. For example,

   choose $7\dfrac{6}{13}, 7\dfrac{5}{13}$, and $7\dfrac{4}{13}$. Then $\dfrac{7\frac{6}{13}}{14} = \dfrac{97}{182}$ and $\dfrac{7\frac{5}{13}}{14} = \dfrac{96}{182}$ and

   $\dfrac{7\frac{4}{13}}{14} = \dfrac{95}{182}$ are between the given fractions.

   c. To obtain fractions between $\dfrac{6}{8}$ and $\dfrac{7}{8}$, choose numerators between 6 and 7, say

   $6\dfrac{1}{3}, 6\dfrac{1}{4}$ and $6\dfrac{1}{5}$. Then $\dfrac{6\frac{1}{3}}{8} = \dfrac{19}{24}, \dfrac{6\frac{1}{4}}{8} = \dfrac{25}{32}$ and $\dfrac{6\frac{1}{5}}{8} = \dfrac{31}{40}$ are between the

   given fractions.

4. a. You will need the larger numbers in the denominators, so make one
      denominator 7 and the other 8. Then put the 6 with the 8 so that you get a
      larger number of smaller pieces.

   $$\dfrac{5}{7} + \dfrac{6}{8}$$

   b. You will need pieces of larger size, so use 5 and 6 as denominators. Make 8
      the numerator for the 5 so that you get a larger number of larger pieces.

   $$\dfrac{8}{5} + \dfrac{7}{6}$$

   c. You want to make the fractions as close as possible in size and you want
      the largest numbers in the denominator so that your difference is very small.
      $\dfrac{5}{8} > \dfrac{6}{7}$ so put 6 as the numerator above 8.

   $$\dfrac{6}{8} - \dfrac{5}{7}$$

SUPPLEMENTARY PROBLEMS

1. Decide which of the two fractions in each pair is larger by using reasoning only. No common denominators or cross multiplying strategies are to be used.

    a. $\dfrac{3}{7}, \dfrac{5}{8}$ b. $\dfrac{3}{7}, \dfrac{2}{5}$ c. $\dfrac{4}{9}, \dfrac{9}{11}$ d. $\dfrac{2}{5}, \dfrac{5}{9}$

    e. $\dfrac{3}{8}, \dfrac{5}{9}$ f. $\dfrac{3}{7}, \dfrac{5}{12}$ g. $\dfrac{6}{11}, \dfrac{7}{12}$

2. Find 3 fractions between each of the following pairs.

    a. $1, 1\dfrac{1}{4}$ b. $\dfrac{1}{2}, \dfrac{3}{4}$ c. $\dfrac{7}{9}, \dfrac{8}{9}$ d. $\dfrac{4}{5}, 1$ e. $\dfrac{1}{3}, \dfrac{2}{3}$ f. $\dfrac{7}{6}, \dfrac{8}{6}$

3. A third grade student, who was shading to compare $\dfrac{3}{8}$ and $\dfrac{4}{9}$, observed that he could use this drawing to add the fractions $\dfrac{3}{8}$ and $\dfrac{4}{9}$. He said he would add up all the shaded blocks, and subtract off the ones that have two kinds of shading because otherwise, they would get counted twice. What do you think about his strategy?

    $\dfrac{3}{8} + \dfrac{4}{9} =$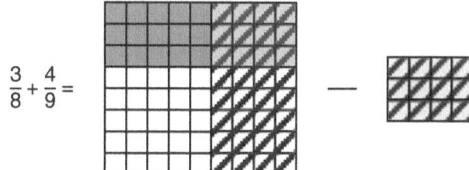

4. Answer the following *Can You See* questions using this figure.

    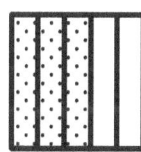

    a. Can you see $\dfrac{3}{5}$ of something?   b. Can you see $\dfrac{5}{3}$ of something?

    c. Can you see $\dfrac{2}{3}$ of $\dfrac{3}{5}$?   d. Can you see $\dfrac{5}{3}$ of $\dfrac{3}{5}$?

    e. Can you see $1 \div \dfrac{3}{5}$?   f. Can you see $\dfrac{3}{5} \div 2$?

    g. Can you see $\dfrac{5}{4} \div \dfrac{3}{4}$?

5. Use Martin's method to find three fractions between

    a. $\dfrac{1}{8}$ and $\dfrac{1}{7}$ b. $\dfrac{2}{5}$ and 1 c. $1\dfrac{3}{7}$ and $1\dfrac{1}{2}$

6. Find 3 equally spaced fractions between

   a. 1 and $1\frac{1}{5}$ b. $\frac{3}{7}$ and $\frac{6}{7}$ c. $2\frac{1}{3}$ and $2\frac{3}{4}$

7. Using each of the numbers 2, 3, 4, and 12 only once, construct a fraction problem with the given properties:

   a. the smallest possible sum
   b. the largest possible sum
   c. the smallest possible positive difference
   d. the largest possible positive difference
   e. the smallest possible product
   f. the largest possible product
   g. the smallest possible quotient
   h. the largest possible quotient

8. Compare by shading. Use a piece of graph paper.

   a. $\frac{2}{3}, \frac{3}{7}$ b. $\frac{1}{4}, \frac{2}{5}$ c. $\frac{4}{9}, \frac{9}{8}$ d. $\frac{7}{9}, \frac{12}{13}$

9. Compare using unit fractions:

   a. $\frac{5}{8}, \frac{3}{4}$ b. $\frac{4}{7}, \frac{5}{6}$ c. $\frac{9}{10}, \frac{5}{8}$ d. $\frac{2}{3}, \frac{5}{9}$ e. $\frac{5}{9}, \frac{7}{15}$

10. Find 3 fractions between $\frac{1}{7}$ and $\frac{1}{6}$.

11. Find 3 fractions between $\frac{3}{5}$ and $\frac{4}{5}$.

12. Compare these fractions using any of the reasoning methods you have encountered so far.

   a. $\frac{2}{3}, \frac{2}{5}$ b. $\frac{6}{7}, \frac{7}{8}$ c. $\frac{5}{9}, \frac{3}{4}$ d. $\frac{1}{5}, \frac{1}{7}$ e. $\frac{9}{13}, \frac{12}{17}$

   f. $\frac{5}{11}, \frac{9}{16}$ g. $\frac{1}{12}, \frac{2}{13}$ h. $\frac{4}{7}, \frac{3}{8}$ i. $\frac{4}{9}, \frac{3}{8}$ j. $\frac{7}{8}, \frac{9}{11}$

PRAXIS QUESTIONS

1. Which is farthest from $\frac{1}{2}$ on a number line?

   a. $\frac{1}{12}$ b. $\frac{7}{8}$ c. $\frac{3}{4}$ d. $\frac{2}{3}$ e. 2

2. Which of the following fractions is more than $\frac{3}{4}$?

a. $\dfrac{35}{71}$ b. $\dfrac{13}{20}$ c. $\dfrac{71}{101}$ d. $\dfrac{19}{24}$ e. $\dfrac{15}{20}$

3. $\dfrac{4}{9}$ is less than which of the following?

   a. 44%   b. 0.45   c. 0.4444   d. $\dfrac{4}{10}$ e. 44.4%

4. All of the following numbers are equal except for:

   a. $\dfrac{4}{9}$ b. $\dfrac{44}{90}$ c. $\dfrac{404}{909}$ d. $\dfrac{444}{999}$ e. $\dfrac{4044}{9099}$

5. Of the following, which answer is not a fractional equivalent to $\dfrac{2}{3}$?

   a. $\dfrac{10}{15}$ b. $\dfrac{12}{18}$ c. $\dfrac{20}{30}$ d. $\dfrac{8}{9}$ e. $\dfrac{40}{60}$

6. If $x < \dfrac{3}{5}$, which of the following could be a value of x?

   a. $\dfrac{3}{6}$ b. $\dfrac{4}{7}$ c. $\dfrac{5}{7}$ d. $\dfrac{18}{57}$ e. $\dfrac{5}{9}$

7. Which fraction is the greatest?

   a. $\dfrac{5}{4}$ b. $\dfrac{99}{100}$ c. $\dfrac{25}{24}$ d. $\dfrac{12}{13}$ e. $\dfrac{17}{16}$

8. Which of the following fractions is equivalent to $\dfrac{5}{6}$?

   a. $\dfrac{10}{18}$ b. $\dfrac{25}{30}$ c. $\dfrac{20}{30}$ d. $\dfrac{15}{24}$ e. $\dfrac{10}{24}$

9. Select the list in which the terms are given in order from least to greatest value:

   a. $0, \dfrac{1}{8}, 0.045, 0.711$ b. $\dfrac{1}{3}, 0.375, 2, \dfrac{14}{5}$ c. $\dfrac{16}{4}, 3.78, 1\dfrac{3}{8}, \dfrac{8}{16}$

   d. $0.441, 0.474, \dfrac{5}{6}, \dfrac{4}{5}$ e. $-\dfrac{2}{3}, 0, \dfrac{1}{3}, \dfrac{4}{5}, \dfrac{2}{3}$

10. Which of the following is greater than $\dfrac{1}{3}$?

    a. 0.33   b. $\left(\dfrac{1}{3}\right)^2$ c. $\dfrac{1}{4}$ d. $\dfrac{1}{0.3}$ e. $\dfrac{0.3}{2}$

11. Which of the following fractions has the least value?

    a. $\dfrac{46}{30}$ b. $\dfrac{241}{159}$ c. $\dfrac{240}{97}$ d. $\dfrac{195}{97}$ e. $\dfrac{56}{42}$

12. When the fractions $\frac{2}{3}, \frac{5}{7}, \frac{8}{11}$, and $\frac{9}{13}$ are arranged in ascending order of size, the result is

   a. $\frac{8}{11}, \frac{5}{7}, \frac{9}{13}, \frac{2}{3}$ b. $\frac{5}{7}, \frac{8}{11}, \frac{2}{3}, \frac{9}{13}$ c. $\frac{2}{3}, \frac{8}{11}, \frac{5}{7}, \frac{9}{13}$

   d. $\frac{2}{3}, \frac{9}{13}, \frac{5}{7}, \frac{8}{11}$ e. $\frac{9}{13}, \frac{2}{3}, \frac{8}{11}, \frac{5}{7}$

13. Which fraction is farthest from 1?

   a. $\frac{11}{12}$ b. $\frac{14}{13}$ c. $\frac{9}{10}$ d. $1\frac{3}{25}$ e. $1\frac{2}{15}$

14. Which fraction sum is between $\frac{1}{4}$ and $\frac{1}{5}$?

   a. $\frac{1}{5}+\frac{1}{24}$ b. $\frac{1}{4}+\frac{1}{20}$ c. $\frac{1}{5}+\frac{1}{3}$ d. $\frac{1}{4}+\frac{1}{10}$ e. $\frac{1}{5}+\frac{1}{4}$

15. Which sum of unit fractions has the smallest value?

   a. $\frac{1}{5}+\frac{1}{3}+\frac{1}{5}+\frac{1}{3}$ b. $\frac{1}{10}+\frac{1}{3}+\frac{1}{9}+\frac{1}{3}$ c. $\frac{1}{7}+\frac{1}{7}+\frac{1}{10}+\frac{1}{10}$

   d. $\frac{1}{3}+\frac{1}{3}+\frac{1}{9}+\frac{1}{9}$ e. $\frac{1}{2}+\frac{1}{3}+\frac{1}{4}+\frac{1}{5}$

## CHAPTER 7

Fractions as Part–Whole Comparisons

DISCUSSION OF ACTIVITIES

1. a. The triangles are 4 of the 9 items in the group: $\dfrac{4}{9}$
    b. The unit is the entire group of figures.
    c. Two triangles are half the set of four triangle: $\dfrac{2}{4} = \dfrac{1}{2}$
    d. In this case, the question makes it clear that the unit is the set of 4 triangles, not the entire set of figures.
    e. Three circles compared to the number of circles in the set is $\dfrac{3}{5}$.
    f. The set of 5 circles is explicitly named as the unit.

2. a. 2 sections = $\dfrac{1}{3}$, 6 sections = 1; so 2 circles = 1
    b. 1 circle = 1
    c. 5 sections = $\dfrac{1}{3}$, 15 sections = 1; so $2\dfrac{1}{2}$ circles = 1
    d. Subdivide the given circle into eighths. 6 sections = $\dfrac{2}{9}$. 3 sections = $\dfrac{1}{9}$, 27 sections = 1; so $3\dfrac{3}{8}$ circles = 1
    e. 4 circles = 1
    f. Subdivide the given circles into twelfths. 14 sections = $\dfrac{2}{3}$, 7 sections = $\dfrac{1}{3}$, 21 sections – 1; so $1\dfrac{3}{4}$ circles = 1
    g. Subdivide the given circles into sixths. 10 sections = $\dfrac{2}{3}$, 5 sections = $\dfrac{1}{3}$, 15 sections = 1; so $2\dfrac{1}{2}$ circles = 1

3. a. Color the bottom half plus the medium rectangle on the top half.
    b. Color the top half plus 9 small squares on the bottom half plus 4 medium squares on the bottom half.
    c. Color the large square on the bottom half plus 3 small squares on the bottom half.
    d. Color the bottom half plus 2 small rectangles on the upper half.
    e. On the bottom half, color the 9 small squares and the large square.
    f. Color the entire upper half and 2 medium squares on the bottom half.

g. Color the large square and the 4 medium squares on the bottom half plus 3 of the small squares on the bottom half.

h. Color 3 small rectangles on the top half.

4. a. $\frac{3}{5}$(5 days)　　　　b. $\frac{12}{9}$(pair)　　　　c. $\frac{1}{12}$(pair)

d. $\frac{2}{3}$(12-pack)　　　　e. $1\frac{1}{3}$(6-packs)　　　　f. $1\frac{1}{2}$(half dollars)

g. $4\frac{1}{4}$(acres)　　　　h. $8\frac{1}{2}$(half acres)

5. a.

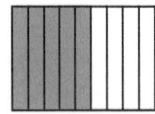

b.

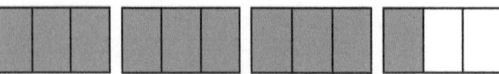

4 acres $= 12\left(\frac{1}{3}\text{-acres}\right)$ and 6 equal parts each contain 2 of the $\frac{1}{3}$ pieces. Shade

5 out of 6 of the $\left(\frac{2}{3}\text{-acre}\right)$ pieces.

c.

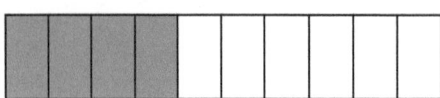

10 acres $= 5$(pair of acres). Shade 2 out of the 5 (pairs of acres).

d.

2 cakes $= 6\left(\frac{1}{3}\text{-cakes}\right)$. Shade 5 of the 6 $\left(\frac{1}{3}\text{-cakes}\right)$.

6. The area of the square is $A = s^2$ or 25 sq. ft. The area of the circle is $A = \pi r^2$ or 4π sq. ft. The part of the square covered by the circle is $\frac{4\pi}{25} = 0.5026$ or just a little over half the square.

7. $\frac{4}{5}$(parts) $= \frac{24}{30}\left(\frac{1}{6}\text{-parts}\right)$ and $\frac{5}{6}$(parts) $= \frac{25}{30}\left(\frac{1}{5}\text{-parts}\right)$ so $\frac{5}{6}$ is larger by $\frac{1}{30}$ of an acre.

On the fraction strips, partition each $\frac{1}{5}$-acre into 6 equal parts and partition each $\frac{1}{6}$-acre into 5 equal parts.

8. Note that in this problem, the unit is the set of 18 stars. Think about what each fraction means in relation to that unit.

   a. 18 stars $= 3$(6-packs). Therefore, $\frac{2}{3}$ is 2 out of 3 (6-packs) or 12 stars.

   b. 18 stars $= 6$ (3-packs). $\frac{5}{6}$ is 5 of 6 (3-packs) or 15 stars.

   c. 18 stars $= 9$(pair). $\frac{7}{9}$ is 7 out of 9 (pair) or 14 stars.

   d. 18 stars $= 12\left(1\frac{1}{2}\text{-packs}\right)$. $\frac{7}{12}$ is 7 out of $12\left(1\frac{1}{2}\text{-packs}\right)$ or $10\frac{1}{2}$ stars.

   e. $\frac{11}{18}$ is 11 out of 18 single stars or 11 stars.

9. There are 6 sets of 3 hearts each. Five of those sets will be $\frac{5}{6}$. So $\frac{5}{6} = 15$ hearts.

   Two sets of 3 taken together make $\frac{1}{3}$ of the hearts. So $\frac{2}{3} = 12$ hearts.

   A pair of hearts $= \frac{1}{9}$. So $\frac{5}{9} = 10$ hearts.

   Arranged smallest to largest, the fractions are: $\frac{5}{9}, \frac{2}{3}, \frac{5}{6}$.

10. Look at packs of 5 squares. $\frac{8}{8}$(5-packs) $= 40$, so $\frac{7}{8}$(5-packs) $= 35$ squares.

11. In this problem, a teacher was trying to find out what her fourth graders thought about halves of different units. Were they absolutely or relatively the same amounts? Mike seems to be using a ratio interpretation, rather than a part–whole comparison. From his answer, it is difficult to tell, but he seems to be saying that the absolute amounts cannot be the same because they come from different units. Adam had some notion that $\frac{1}{2}$ describes both pictures, and tried to show that the areas are not the same amount in absolute terms. Derek identified the appropriate part–whole fraction in each picture, but answered a different question than he was asked. He seemed to be saying that both of his fractions were equivalent to $\frac{1}{2}$, but he gave no indication of what he thought about absolute or relative amounts. These children were quite young and part of the problem—even if they *do* understand—is knowing how to talk about these concepts. The teacher did not get as much information as she wanted. The question was difficult for fourth graders and it did not occur to most of them to mention that the halves referred to different units. The teacher took it from the class responses that most of them knew that the fractions did not represent the same absolute amounts.

12. a. Partition so that there are 6 equal pieces in each cake.
     b. 12 equal pieces in each cake
     c. 9 equal pieces in each cake
     d. 12 equal pieces in each cake
     e. 20 equal pieces in each cake
     f. 24 equal pieces in each cake

13. a. Partition each fraction strip so that ther are 15 equal pieces in each. Then you can see that $\frac{2}{3}$ of an acre is larger by $\frac{1}{15}$ of an acre.

     b. Partition the models so that there are 18 equal pieces in each. Then, $\frac{5}{6}$ of a mile is larger by $\frac{1}{18}$ of a mile.

     c. Partition each fraction strip so that there are 20 equal pieces in each. Then $\frac{3}{4}$ of a cheese cake is larger by $\frac{1}{20}$.

     d. $\frac{7}{8}$ is $\frac{1}{24}$ larger than $\frac{5}{6}$.

     e. $\frac{7}{9}$ of a pizza is $\frac{1}{9}$ larger than $\frac{2}{3}$ of a pizza

     f. $\frac{5}{6}$ of a cake is $\frac{1}{12}$ larger than $\frac{3}{4}$ of a cake

14. Partitioning the cakes so that each has 12 equal pieces, you can see that Maurice has $2\frac{8}{12}$ and Sam has $3\frac{3}{12}$. Together they have $2\frac{8}{12} + 3\frac{3}{12} = 5\frac{11}{12}$ cakes.

15. What is $1\frac{2}{9} \div 1\frac{2}{3}$? Ask yourself: How many copies of $1\frac{2}{3}$ can I measure out of $1\frac{2}{9}$? Partition thirds so that there are 9 equal pieces. Then $1\frac{2}{3}$ is 15 of those equal pieces. $1\frac{2}{9}$ is 11 of those equal pieces. How many times can you measure 15 parts out of 11 parts? $\frac{11}{15}$ (less than once).

16. What is $2\frac{1}{3} \div 1\frac{1}{4}$? How many copies of $1\frac{1}{4}$ can I measure out of $2\frac{1}{3}$? Partition a third strip and a fourth strip so that each has 12 equal pieces. Then $2\frac{1}{3}$ is composed of 28 of those pieces and $1\frac{1}{4}$ is composed of 15 of those pieces. How many times can you measure 15 parts out of 28 parts? $1\frac{13}{15}$ times.

17. a.

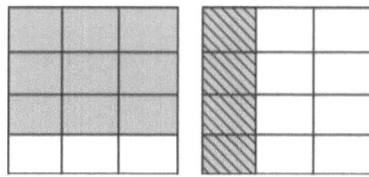

Remember that the unit is 1 rectangle and it consists of 12 pieces. In the model, 12 pieces are both shaded and colored, so the answer is 1.

b.

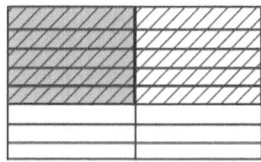

There are 5 hatched and colored pieces out of the 16 pieces in the unit area, so the answer is $\frac{5}{16}$.

c.

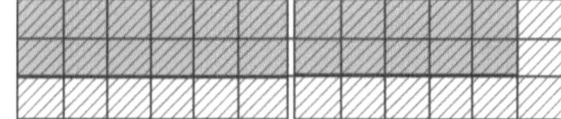

22 pieces are shaded and there are 18 in the unit. This means that the answer is $\frac{22}{18} = 1\frac{4}{18}$.

18. Represent $1\frac{3}{4}$ and partition so that you get twelfths. $\frac{2}{3} = \frac{8}{12}$ and $1\frac{3}{4} = \frac{21}{12}$. How many times can you measure 8 out of 21? 2 times with 5 pieces left over. Remember that in a division problem, the divisor is the unit. So the remainder $\frac{5}{8}$.

19. a. A person in group B.   b. A person in group B.
     c. A person in group A.   d. A person in group A.
     e. A person in group A.

20. a. $3\frac{1}{2}$ boxes   b. 6 boxes   c. $3\frac{1}{3}$ boxes

   d. 10 boxes   e. $3\frac{1}{3}$ boxes

21. All of the children were correct. T.M.P. and Tracy both used the same strategy. They lined up $\frac{1}{3}$ of the unit with $\frac{1}{6}$ of the unit, translated the $\frac{1}{3}$ into $\frac{2}{6}$, and named the sum $\frac{3}{6}$. Kyle used a more sophisticated strategy. He put 3 of the $\frac{1}{3}$ pieces together to make a unit, then he shaded $\frac{1}{3}$ and $\frac{1}{6}$ of the unit. Having superimposed the two addends on the unit, he could see that they were half of the unit. Addy aligned the $\frac{1}{3}$ piece with the $\frac{1}{6}$ piece and then introduced the $\frac{1}{12}$ piece. This was an extraneous step. She failed to notice that she could name the sum in terms of sixths and that a larger denominator was not necessary. Nevertheless, her answer was correct.

22. It is a common misconception among young children that when something is cut into two pieces those pieces are called halves. It is incorrect to call a piece a half unless it is one 2 equal pieces.

23. The friends had already partitioned the candy into thirds, but when the new person came, they needed 4 equal shares. Each of the first three boys should fold his licorice into 4 equal sections and cut off one section for the fourth boy. That way, everyone ends up with $\frac{3}{12}$ of the licorice.

24. a. Shade 2 columns.
   b. Shade 13 pairs of small squares.
   c. Shade 2 groups consisting of 4 small squares each.
   d. Shade 7 groups consisting of 3 small squares each.
   e. Shade 7 groups consisting of 4 small squares each.
   f. Shade $\frac{1}{3}$ and $\frac{1}{4}$ in the same unit to get $\frac{7}{12}$.
   g. $\frac{7}{18}$ h. $\frac{8}{9}$ i. $\frac{5}{12}$ j. $\frac{7}{12}$

25. These children had developed multiple interpretations for fractional numbers and had a good sense of the way things "work" in the fraction world. Several of them showed the power of unitizing. Grace knew that $\frac{2}{3}$ is twice the size of $\frac{1}{3}$. She also

knew that if she measured with a piece that was twice as big, she would get only half as many copies of it. Essentially, she told us that if there are three $\frac{1}{3}$s in 1, then there can only be one and half copies of $\frac{2}{3}$ in 1. Lindsay, as well, thought in chunks of size $\left(\frac{2}{3}\text{-unit}\right)$. Troy thought of both the whole and the divisor in chunks of size $\left(\frac{1}{6}\text{-unit}\right)$, then measured $\frac{1}{6}$-units out of the whole. Underlying Carson's strategy was an understanding of a fraction as a quotient. He understood $\frac{2}{3}$ as the result of dividing 2 units into 3 equal shares. He then claimed that starting with only 1 rectangle would yield half as much.

SUPPLEMENTARY PROBLEMS

1. 10 cents is what part of a dollar? Find 3 different ways to name the part of a dollar.

2. Use this set of diamonds to order these fractions from smallest to largest:
    $$\frac{7}{9}, \frac{4}{3}, \frac{5}{9}, \frac{1}{2}$$

3. Using the methods of this chapter, draw pictures to show each quantity.
    a. I had 3 cakes, and $\frac{5}{6}$ of them were eaten. How much cake was eaten?
    b. I own 5 acres and $\frac{7}{15}$ of my property is covered by a lake. How many acres are covered by water?
    c. I had 32 hard candies, but I have only $\frac{3}{8}$ of them left. How many are left?
    d. I have 2 acres of land and $\frac{3}{4}$ of it is wooded. How many acres are wooded?

4. Draw an appropriate picture to show each operation and write the division statement symbolically $(a \div b = c)$.
    a. How many $\frac{1}{2}$'s are there are in 8?
    b. How many times you can measure $\frac{2}{3}$ out of 3?

5. Two students responded to this question: Which is more, $\frac{3}{5}$ of a pie or $\frac{7}{11}$ of a pie. Are their answers correct?

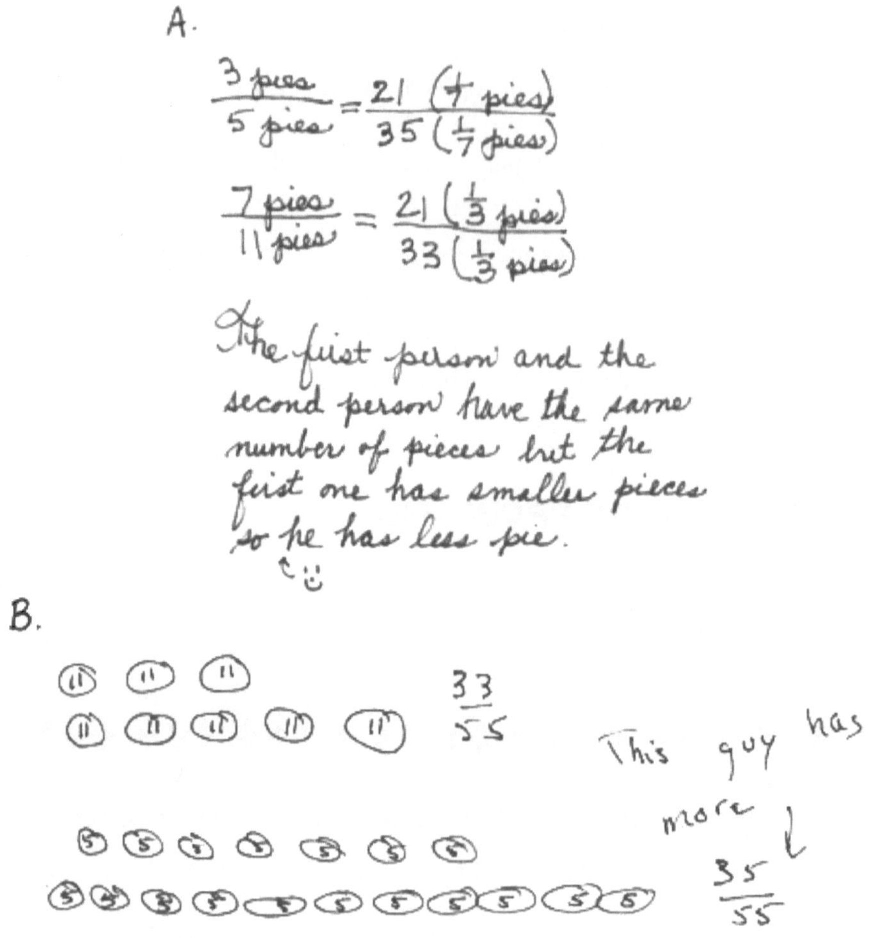

A.

$$\frac{3 \text{ pies}}{5 \text{ pies}} = \frac{21 \left(\frac{1}{7} \text{ pies}\right)}{35 \left(\frac{1}{7} \text{ pies}\right)}$$

$$\frac{7 \text{ pies}}{11 \text{ pies}} = \frac{21 \left(\frac{1}{3} \text{ pies}\right)}{33 \left(\frac{1}{3} \text{ pies}\right)}$$

The first person and the second person have the same number of pieces but the first one has smaller pieces so he has less pie. ♥☺

B.

$\dfrac{33}{55}$

This guy has more ↓

$\dfrac{35}{55}$

7. Solve by unitizing.

a. Which is larger, $\dfrac{5}{6}$ of a cake or $\dfrac{2}{3}$ of a cake?

b. $\dfrac{4}{5}$ of a pizza or $\dfrac{5}{6}$ of a pizza?

c. $\dfrac{2}{3}$ of an acre or $\dfrac{3}{5}$ of an acre?

d. $\dfrac{5}{6}$ of a mile or $\dfrac{7}{9}$ of a mile?

e. $\dfrac{3}{4}$ of a cherry pie or $\dfrac{7}{10}$ of the same pie?

8. Two candles of equal length are lighted at the same time. One candle take 9 hours to burn out and the other takes 6 hours to burn out. After how much time will the slower-burning candle be exactly twice as long as the faster-burning candle?

9. Jim gave half the money in his pocket to Chris. He gave a fourth of what was left to Tom. then he gave a third of what was left to Zoe. Then he split the remainder with you. If Jim gave you $20, how much money did he have before he started giving it away?

10. In each case, use the fraction strips in *MORE* to tell which fractions is larger. How much larger?

a. $\frac{2}{3}$ or $\frac{1}{2}$ b. $\frac{1}{3}$ or $\frac{2}{5}$ c. $\frac{5}{6}$ or $\frac{2}{3}$ d. $\frac{1}{4}$ or $\frac{2}{9}$

e. $\frac{2}{3}$ or $\frac{5}{8}$ f. $\frac{5}{8}$ or $\frac{9}{12}$ g. $\frac{5}{6}$ or $\frac{7}{8}$

11. Shade the amount indicated above each unit area and use unitizing to generate equivalent names for the shaded amount.

a. $\frac{1}{2}$ b. $\frac{1}{4}$ c. $\frac{2}{3}$

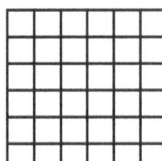   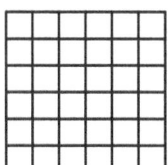   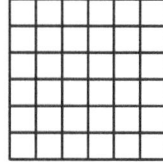

d. $\frac{3}{4}$ e. $\frac{5}{9}$ f. $\frac{1}{6}$

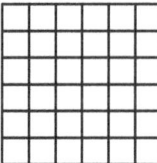   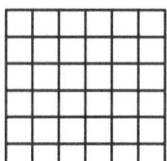   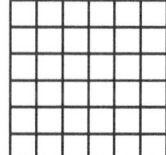

g. $\frac{3}{9}$ h. $\frac{11}{18}$ i. $\frac{5}{6}$

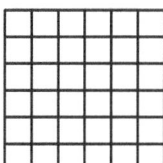   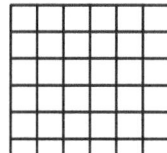   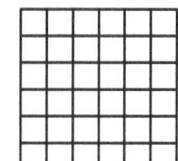

j. $\dfrac{5}{12}$

k. $\dfrac{1}{3} + \dfrac{1}{6}$

l. $\dfrac{1}{9} + \dfrac{1}{12}$

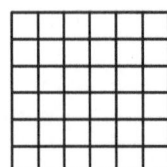   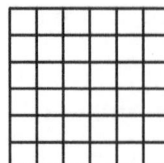   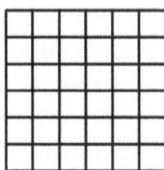

m. $\dfrac{1}{2} + \dfrac{1}{3}$

n. $\dfrac{2}{3} + \dfrac{1}{9}$

o. $\dfrac{1}{4} + \dfrac{7}{12}$

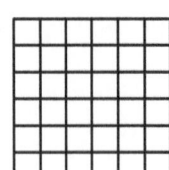   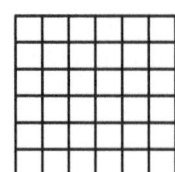   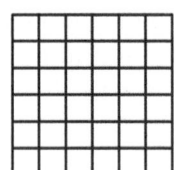

12. JS answered the question: which is larger, $\dfrac{4}{6}$ or $\dfrac{7}{9}$. Is he correct?

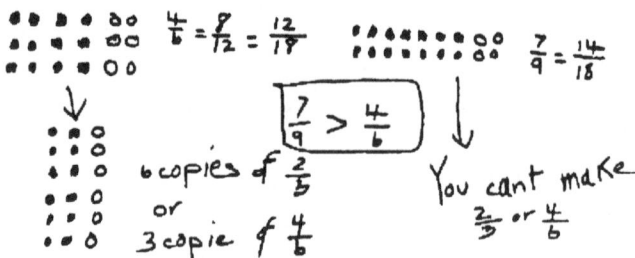

PRAXIS QUESTIONS

1. The students at Rosewood High School travel to school in five different ways. $\dfrac{3}{8}$ of them are driven by car; $\dfrac{1}{3}$ take the yellow school bus; $\dfrac{2}{9}$ of them ride their bicycles; $\dfrac{1}{24}$ use the Metro bus; the rest walk. What fraction of the students walk?

a. $\dfrac{7}{44}$ b. $\dfrac{35}{36}$ c. $\dfrac{1}{36}$ d. $\dfrac{1}{18}$ e. none of these

2. If a cake recipe calls for $2\frac{1}{2}$ cups of flour and you make 3 of these cakes, the number of cups of flour you will need is

    a. 5   b. $6\frac{1}{2}$ c. $7\frac{1}{2}$ d. 9   e. $9\frac{1}{2}$

3. What is the average of $\frac{1}{2}, \frac{2}{3}$, and $\frac{15}{12}$?

    a. $\frac{29}{12}$ b. $\frac{29}{24}$ c. $\frac{29}{52}$ d. $\frac{29}{44}$ e. $\frac{29}{36}$

4. Rob has gone $3\frac{1}{5}$ miles of his 5-mile run. How many more miles has he left to run?

    a. a little less than 2 miles   b. a little more than 2 miles
    c. a little less than 3 miles   d. a little more than 3 miles
    e. a little less than 4 miles

5. The product $\frac{3}{4}(\frac{1}{2}\cdot\frac{2}{3})$ in reduced form is

    a. $\frac{1}{4}$ b. $\frac{3}{4}$ c. $\frac{6}{24}$ d. $\frac{1}{2}$ e. $\frac{3}{12}$

6. 7.17 is between

    a. 7.0 and 7.2   b. 7.02 and 7.10   c. 7.5 and 7.9
    d. 7.00 and 7.04   e. 7.012 and 7.102

7. One-tenth is what part of three fourths?

    a. $\frac{40}{3}$ b. $\frac{3}{40}$ c. $\frac{15}{2}$ d. $\frac{1}{8}$ e. $\frac{2}{15}$

8. Estimate the answer for $124 \cdot \frac{49}{24}$

    a. 200   b. 250   c. 325   d. 500   e. 490

9. A punch recipe calls for $3\frac{3}{4}$ cups of pineapple juice for 10 servings. How many cups of pineapple juice should be used to make 5 servings of the same punch?

    a. 2   b. $1\frac{3}{8}$ c. $1\frac{7}{8}$ d. $1\frac{3}{4}$ e. none of these

10. Tameka runs $\frac{2}{3}$ of a mile each morning. Today she stopped running after going $\frac{4}{5}$ of the way. How far did she run?

    a. $\frac{2}{15}$ b. $\frac{2}{5}$ c. $\frac{3}{4}$ d. $\frac{8}{15}$ e. none of these

11. If $\frac{2}{3}$ of a set consists of 8 items, $\frac{5}{6}$ of the set consists of how many items?

   a. 10   b. 14   c. 12   d. 8   e. none of these

12. Which value of a will make the fraction $\frac{1}{a}$ greater than 1?

   a. 10   b. 1   c. 0   d. $\frac{1}{4}$ e. none of these

13. The winner of a race received $\frac{1}{3}$ of the total purse. The third place finisher received $\frac{1}{3}$ of the winner's share. If the third place finisher won $900, what was the total purse?

   a. $3,000   b. $8,100   c. $900   d. $2,700   e. $1,800

14. A farmer has ten acres of land. Corn is planted on $\frac{5}{6}$ of his land. How many acres are not planted with corn?

   a. $8\frac{2}{3}$ b. $1\frac{2}{3}$ c. $8\frac{1}{3}$ d. $1\frac{1}{3}$ e. $\frac{1}{6}$

15. Which pair of fractions are equivalent?

   a. $\frac{8}{18},\frac{1}{3}$ b. $\frac{4}{6},\frac{8}{24}$ c. $\frac{0}{6},\frac{0}{12}$ d. $\frac{24}{28},\frac{3}{7}$ e. none of these

CHAPTER 8

Fractions as Quotients

DISCUSSION OF ACTIVITIES

1. The children all proposed to give each person $1\frac{1}{3}$ candy bars, but they also proposed different ways of breaking the candy bars. In each case, one share of the candy looks different and is named by different fractions:

$$\frac{1}{3}+\frac{1}{3}+\frac{1}{3}+\frac{1}{3}$$
$$\frac{1}{2}+\frac{1}{2}+\frac{2}{6}$$
$$\frac{1}{2}+\frac{1}{2}+\frac{1}{3}$$
$$\frac{2}{6}+\frac{2}{6}+\frac{2}{6}+\frac{2}{6}$$

Based on these partitions, the following fractions must be equivalent:

$$\frac{2}{6}=\frac{1}{3}$$
$$\frac{4}{3}=\frac{8}{6}$$
$$\frac{4}{3}=\frac{8}{6}=1\frac{2}{6}=1\frac{1}{3}$$

2. Each person will eat $\frac{4}{5}$ of a candy bar, which is $\frac{1}{5}$ of the total candy.

3. One share will be $\frac{2}{4}$ or $\frac{1}{2}$ of a 6-pack. Each share is $\frac{1}{4}$ of the unit.

4. Student B used the least sophisticated strategy. The student split apart every 6-pack to make individual cans and then distributed them to the 3 children. Student C used the most sophisticated strategy because he or she used the least amount of cutting (separating) and marking. This student merely distributed entire 6-packs as far as possible, then distributed individual cans as necessary. Student A's strategy is between the other two in sophistication. Although Student A distributed 6-packs, he or she needed to mark the individual components of the 6-packs. Notice how the student drew each individual can within the 6-pack. This need to mark individual components of a composite unit is usually a sign that a student is in a transitional stage: he or she is beginning to think in terms of

composite units, but still needs some visual reassurance that each person is getting the same number of cans.

5. Figure out the number of cuts that it would take to accomplish each of the partitions. Student A's strategy requires 6 cuts; B's requires 4 cuts; C's requires 10 cuts; and D's requires 8 cuts. You can see that Students A and B anticipated that a share would consist of 4 small rectangles, and they made some effort to keep each share connected. Students C and D were unable to anticipate that a share would consist of 4 rectangles, so they started by distributing 1 or 2 at a time. Ranking by sophistications (connectedness of shares), we get B, A, D, and C, from highest to lowest.

6. a. At table A, each child gets 1 cookie. At table B, each child gets $\frac{5}{4}$ cookies or $1\frac{1}{4}$ cookies. A child at table B gets $\frac{1}{4}$ of a cookie more than a child at table A.

   b. At table A, each child gets $\frac{7}{3}$ cookies or $2\frac{1}{3}$ cookies. At table B, each child gets 2 cookies. A child at table A gets $\frac{1}{3}$ of a cookie more than a child at table B.

   c. At table A, each child gets $\frac{8}{3}$ cookies or $2\frac{2}{3}$ cookies. At table B, each child gets $\frac{10}{4}$ cookies or $2\frac{1}{2}$ cookies. A child at table A gets $\frac{1}{6}$ of a cookie more than a child at table B.

   d. This is fair.

   e. At table A, each child gets $\frac{1}{3}$ of a cookie. At table B, each child gets $\frac{1}{4}$ of a cookie. A child at table A gets $\frac{1}{12}$ of a cookie more than a child at table B.

   f. At table A, each child gets $\frac{4}{3}$ or $1\frac{1}{3}$ cookies. At table B, each child gets $\frac{5}{4}$ cookies or $1\frac{1}{4}$ cookies. A child at table A gets $\frac{1}{12}$ of a cookie more than a child at table B.

   g. At table A, each child gets $\frac{2}{3}$ of a cookie. At table B, each child gets $\frac{5}{4}$ or $1\frac{1}{4}$ cookies. A child at table B gets $\frac{7}{12}$ of a cookie more than a child at table A.

7. a. Each girl gets one whole cake, but each boy gets more than one cake. A boy gets $1\frac{1}{3}$ cakes. A boy gets $\frac{1}{3}$ of a cake more than a girl gets.

   b. The girls get $\frac{2}{3}$ cake each. Suppose the boys get the same. The first 3 boys use 2 cakes and what remains is 1 cake for 2 boys, so those 2 boys each get only half a cake, not $\frac{2}{3}$ cake. This means that overall, the boys have less cake to split and each boy gets less cake. Suppose you give the fourth boy $\frac{2}{3}$ cake, then there is only $\frac{1}{3}$ cake left for the last boy. This means that as a group, the boys are down by $\frac{1}{3}$ cake; individually, they get $\frac{1}{15}$ of a cake less than each girl gets.

c. The boys get 4 cakes for 5 people, or $\frac{4}{5}$ cake each. On the girls' side, assign 4 cakes to the first 5 girls. Then there is 1 whole cake left for the last girl. She gets more than $\frac{4}{5}$ cake. This means that the girls are up by $\frac{1}{5}$ cake. Shared 6 ways, it means that each girl gets $\frac{1}{30}$ of a cake more than a boy gets.

d. The girls get 3 cakes for 5 people. If you apply that ratio on the boys' side, the boys are down by 2 cakes. They each get $\frac{2}{15}$ of a cake less than a girl gets. Conversely, a girl gets $\frac{2}{15}$ of a cake more than a boy gets.

e. Each girl gets $1\frac{1}{2}$ pizzas. Each boy gets 1 pizza. A girl gets $\frac{1}{2}$ of a pizza more than a boy gets.

f. Each girl gets $\frac{1}{2}$ of a cake. If you try to give each boy $\frac{1}{2}$ of a cake, 1 boy doesn't get any. So the boys are down by $\frac{1}{2}$ of a cake, which means that individually, each gets $\frac{1}{14}$ of a cake less than a girl gets. So a girl gets $\frac{1}{14}$ of a than a boy gets.

g. The girls have 2 cakes for 5 people. Applying that ratio on the boys' side leaves the last 4 boys with 1 cake. Two of those boys can get $\frac{2}{5}$ of a cake, but then there are 2 boys left with only $\frac{1}{5}$ of a cake. So the boys are down by $\frac{3}{5}$ cake. Sharing that shortage, each boy gets $\frac{3}{45}$ or $\frac{1}{15}$ of a cake less than each girl gets.

h. Each girl gets $\frac{1}{3}$ of a pizza. If each boy gets $\frac{1}{3}$ of a pizza, they boy have $1\frac{1}{3}$ or $\frac{4}{3}$ pizzas extra. Sharing that extra pizza, each boy gets $\frac{4}{15}$ of a pizza more than a girl gets.

i. The boys get 3 pizzas for 5 people, or $\frac{3}{5}$ of a pizza each. Apply the boy ratio to the girls' side, the last girl has 1 whole pizza. This means that the girls have an extra $\frac{2}{5}$ of a pizza. Shared 6 ways, that means that a girl gets $\frac{2}{30}$ or $\frac{1}{15}$ of a pizza more than a boy gets.

8. Answers will vary.

9.   Answers will vary.

10. Compare the children on the first criterion: number of cuts. Nicole used many cuts, Eric used few, but Steve was most economical. The ability to look at how much stuff you have and make an estimate of how large a chunk each person will receive, as opposed to cutting and distributing pieces to see how far you get, is a more mature approach to partitioning. Also compare the students on their ability to give a single fraction name to one person's share. Nicole, whose shares were most fragmented, did not answer the question *how much*. Eric's shares were less fragmented and he was able to identify $\frac{4}{6}$ of a candy bar as one share. Steve, who had done the least cutting, was able to see that each person would receive $\frac{2}{3}$ of a candy bar.

11. Solution A: Using the girls' ratio as the measure.

   a. Who has more pizza, the girls' group or the boys' group (relatively speaking)? Measure the ratio (2 pizzas : 3 people) out of the boys' side. That is, give each boy $\frac{2}{3}$ of a pizza, the same as a girl's share. Then there are 2 boys left with 2 pizzas, so each gets more than $\frac{2}{3}$ of a pizza. This shows that the boys' group has more pizza.

   b. Who gets more, a girl individual or a boy individual? Because the boys have more pizza as a group (relatively speaking), each boy individual will receive more pizza than each girl individual.

   c. How much more pizza does the boys' group have? After giving each boy $\frac{2}{3}$ of a pizza (the same amount each girl gets), there will be $\frac{2}{3}$ of a pizza left over. So the boys' group has an extra $\frac{2}{3}$ of a pizza.

   d. How much more pizza does an individual boy receive? The extra $\frac{2}{3}$ of a pizza gets divided up among the 5 boys, so each boy gets $\frac{2}{15}$ of a pizza more than an individual girl gets.

   Solution B: Using the boys' ratio as the measure.

   a. Who has more pizza, the girls' group or the boys' group (relatively speaking)? Measure the ratio (4 pizzas : 5 people) out of the girls' side. That is, give each girl $\frac{4}{5}$ of a pizza, the same amount that a boy receives. Then 2 girls get $\frac{4}{5}$ of a pizza, but the third girl gets only $\frac{2}{5}$ of a pizza. The girls' group is short, so the boys' group must have more pizza.

b. Who gets more, a girl individual or a boy individual? Because the boys have more pizza as a group, each boy individual will receive more pizza than a girl individual will receive.

c. How much less pizza does the girls' group have? The girls' group is short $\frac{2}{5}$ of a pizza. When the 3 girls share the loss, $\frac{2}{5}$ of a pizza is divided among the 3 girls, so that each girl gets $\frac{2}{15}$ of a pizza less than each boy gets.

12. In a sharing situation, for example, $\frac{3}{4}$ designates 3 items divided among 4 people. It signifies the operation of division. It is also a number that is the result of that division. When 3 people share 3 items, the amount in each person's share is designated by $\frac{3}{4}$. $\frac{3}{4}$ is also a ratio in a sharing scenario because it compares, say, number of pizzas to number of people sharing the pizzas. It is not a part–whole fraction.

13.

$\boxed{3+4}$ $\longrightarrow$ $\frac{1}{2} + \frac{4}{6} = \frac{1}{2}$ cheese $+ \frac{2}{3}$ veggie
     6

14.

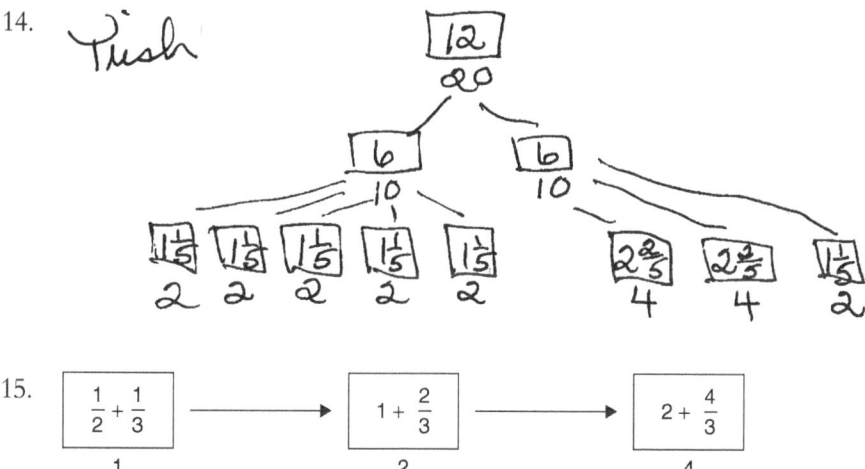

15.

$\boxed{\frac{1}{2} + \frac{1}{3}}$ $\longrightarrow$ $\boxed{1 + \frac{2}{3}}$ $\longrightarrow$ $\boxed{2 + \frac{4}{3}}$
     1   2   4

Four people get more than 2 pizzas, so they cannot be sitting at a table where 5 people share 2 pizzas.

16.   Yes. It is fair. Here is a student solution.

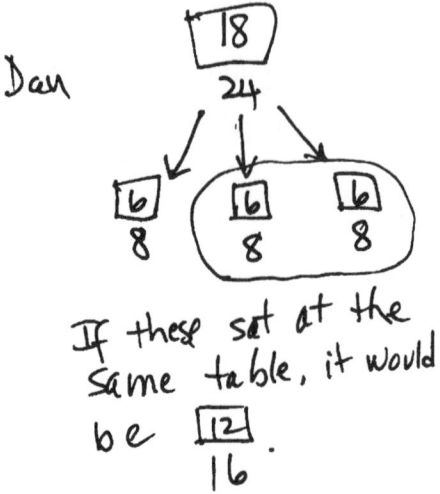

17.   If 5 people shared 2 pizzas, then 30 would share 12. If 5 share 1 cheese pizza, then 30 would share 6. So there must have been 12 veggie pizzas and 6 cheese pizzas.

18. $\boxed{3+4}$ for 8 means each got $\dfrac{3}{8}$ (cheese) + $\dfrac{1}{2}$ (pepperoni)

19.   Four people sharing 1 pizza is like 16 sharing 4. Two people sharing 1 pizza is like 16 sharing 8. So there must have been 4 cheese pizzas and 8 pepperoni pizzas for 16 people.

$$\frac{\boxed{4+8}}{16} \quad = \quad \frac{\boxed{1+2}}{4} \quad = \quad \frac{\boxed{\frac{1}{2}+1}}{2}$$

The 2 people got half a cheese pizza and 1 pepperoni pizza.

SUPPLEMENTARY ACTIVITIES

1.   Shade-to-compare activities can be very ineffective. Analyze the work of this third grade student. Then use your fraction strips to determine which fractional part is greater and by how much.

2. Analyze this student's work. Is it correct?

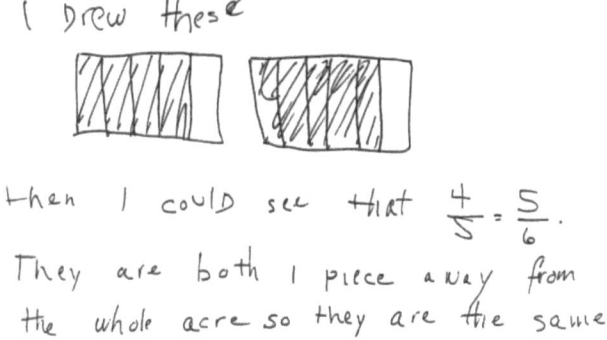

I Drew these

then I could see that $\frac{4}{5} = \frac{5}{6}$.
They are both 1 piece away from
the whole acre so they are the same.

3. For a–c below, decide who gets more pizza, a person seated at the table on the left, or a person seated at the table on the right. How much more pizza will each person at that table receive? Assume that all pizzas are the same size and type.

a.

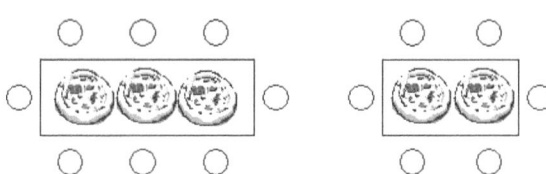

b.

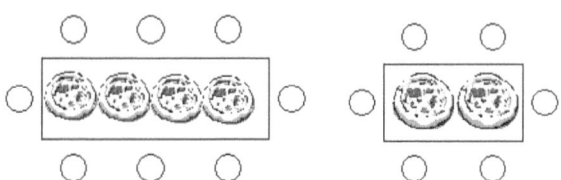

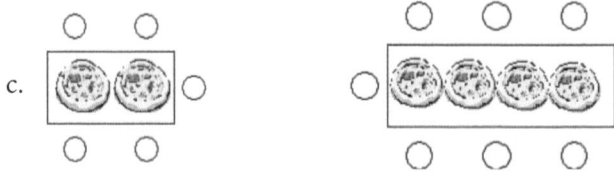

c.

4. Alice and Brad will share a large cookie. Is there a way to partition the cookie so that one child gets two pieces and the other child gets three pieces, but they both get the same amount?

5. If 4 people share 6 candy bars, how much candy will each person receive?

6. If 4 people share 6 candy bars, how much of the total candy will each receive?

7. Three people share the following candy bars. Show three different partitions, write the fractions to represent the pieces in each share, and note the equivalencies.

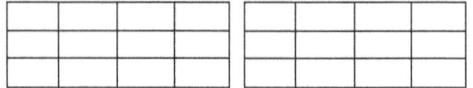

8. If 3 people share these candy bars, how much will one share be? What part of the unit is each share?

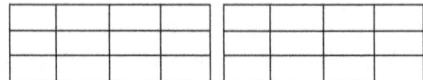

9. Eight people share 6 cheese and 4 mushroom pizzas. How much of each type of pizza is in a share?

10. Suppose three girls share 3 pepperoni pizzas and 4 cheese pizzas. How much will each person get? If each of them takes her portion home and later shares it with her sister, how much will each girl eat?

11. Here are some results you obtained from the children in your class as they considered what one share would be when 4 people shared three identical pizzas. Name each share as a sum of its parts, and list all of the equivalencies that result.

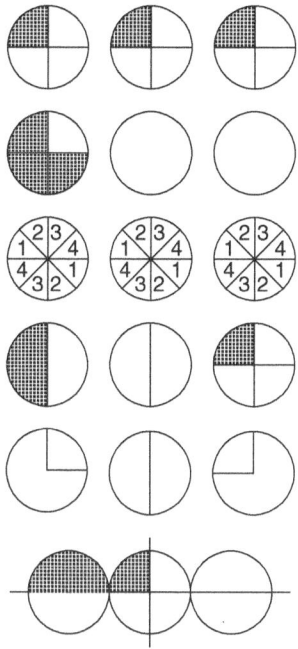

12. Decide who gets more pizza, person A or a person B? Write "same" if everyone gets the same amount. Tell how you know, but do not do any work. All of these questions can be answered by reasoning alone.

a.

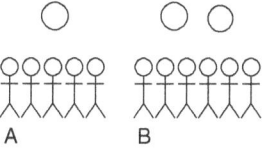

b.

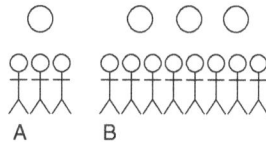

c.

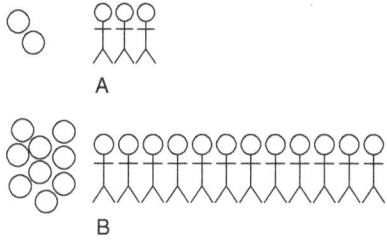

d.

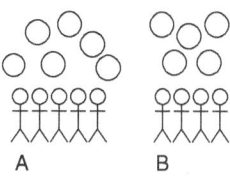

13.   Who gets more, a girl or a boy? How much more?

a.

b.

c.

d.

14.   Twenty four people are going to Maria's Pizzeria for a birthday dinner. Two tables that seat 6 and three tables that seat 4 people have been prepared for this group. If they order 18 pizzas, how should the waiter distribute the pizzas at the 5 tables?

PRAXIS QUESTIONS

1. $\frac{2}{3}$ of a pound of candy will be equally distributed to 6 party bags. Which statement reflects this situation?

    a. $\frac{2}{3} \div \frac{1}{6}$ b. $\frac{2}{3} \cdot \frac{1}{6}$ c. $\frac{2}{3} \cdot 6$ d. $6 \div \frac{2}{3}$ e. $6 \cdot \frac{2}{3}$

2.   6 boys are ordering pizza. If each wants $\frac{4}{6}$ of a pizza, how many pizzas should they order?

    a. 2   b. 3   c. 4   d. 6   e. not given

3.   7 identical candy bars are shared by 5 people. What part of the total candy does each person receive?

    a. $\frac{1}{5}$ b. $\frac{1}{7}$ c. $\frac{7}{5}$ d. $\frac{5}{7}$ e. $1\frac{1}{5}$

4. People are seated at 5 tables, and pizzas have been served to each table. If you wish to get as much pizza as possible, which table would you join?

    a. 6 pizzas, 5 people   b. 5 pizzas, 4 people   c. 8 pizzas, 6 people
    d. 4 pizzas, 3 people   e. 7 pizzas, 5 people

5. A package of macaroni containing $15\frac{1}{2}$ ounces says that the suggested serving is $3\frac{1}{4}$ ounces per person. How many people does it serve?

    a. 2   b. 3   c. 4   d. 5   e. 6

6. How many people shared this candy bar if each person ate $\frac{4}{40}$ of the bar?

    a. 4   b. 20   c. 40   d. 10   e. 5

7. 120 people placed an order for pizza and shared the pizzas fairly. One person's plate contained $\frac{1}{3}$ of a pepperoni pizza, $\frac{1}{6}$ of a cheese pizza, and $\frac{1}{4}$ of a vegee pizza. How many pizzas had the group ordered?

    a. 60   b. 80   c. 100   d. 120   e. none of these

8. Which problem can be modeled by partitioning a segment into smaller segments of equal length?

    a. A coach wanted her runners to do 24-yard relays with 6 team members at each hand-off station. How many stations will there be?
    b. You have 24 feet of ribbon and each package you tie requires 6 feet. How many packages can you decorate?
    c. 24 runners run around a 6-kilometer course. How far does each person run?
    d. If a room is 24 feet long and 6 feet wide, how many sides does it have?
    e. If you drove at 24 mph and you had to travel 6 miles, how many hours would it take you?

9. If 6 children share these candy bars, which of the following does not describe a fair share?

    a. $\frac{2}{3}$ of the candy   b. $\frac{2}{3}$ of a bar   c. $\frac{4}{24}$ of a bar

    d. $\frac{2}{12}$ of each bar   e. $\frac{4}{6}$ of a bar

10.   If 8 people share 2 large identical pizzas, which statement is true?

    a. Each gets $\frac{1}{4}$ of each pizza.   b. Each gets $\frac{1}{8}$ of a pizza.

    c. Each gets $\frac{1}{8}$ of the total pizza.   d. Each gets $\frac{1}{8}$ of each pizza.

    e. Each gets $\frac{1}{4}$ of the total pizza.

11.   If 9 men clocked in a total of 5 hours at the gym, and 7 women clocked in a total of 4 hours, on the average, about how much more time did a woman spend at the gym?

    a. 1 hour   b. $1\frac{2}{3}$ hours   c. 1 minute

    d. $\frac{1}{2}$ of a minute   e. none of these

12.   A large party of 48 people rented a pizza palace for a party and was seated at tables of various sizes. They ordered 30 pizzas and asked the waiters to distribute the pizzas so that every table got a fair share. I observed 5 tables and could see that one of them was short-changed. Which one?

    a. A table of 24 people got 15 pizzas.

    b. A booth for 2 people got $1\frac{1}{4}$ pizzas.

    c. A table of 8 people got 5 pizzas.

    d. A table of 6 people got $3\frac{3}{4}$ pizzas.

    e. A table of 12 people got 7 pizzas.

13.   A group of 5 people share 3 pounds of pasta. Another group of 8 people share 7 pounds of pasta. About how many more pounds of pasta are needed to make sure that both groups get a fair share?

    a. 2   b. 5   c. $\frac{1}{4}$ d. $1\frac{1}{3}$ e. 11

14.   Three partners who won a $1000 Power Ball jackpot decided to share it proportionally according to the number of people in each of their families. The first man won $125 for himself and his wife. The second man is part of a family of 5. How many people are in the third family?

    a. 7   b. 8   c. 9   d. 10   e. none of these

15.   If each person in a large group was served $\frac{2}{3}$ of a cheese pizza and 18 pizzas were ordered, how many people were sharing the order?

    a. 18   b. 27   c. 36   d. 12   e. 24

CHAPTER 9

Fractions as Operators

DISCUSSION OF ACTIVITIES

1. $\frac{3}{12}$ or $\frac{1}{4}$ because $\frac{1}{3} \cdot \frac{3}{4} = \frac{3}{12} = \frac{1}{4}$.

2. a. In each case, the system name is $\dfrac{\text{final output}}{\text{initial input}}$. You should get

$1, \dfrac{2}{20} = \dfrac{1}{10}, \dfrac{9}{2}, \dfrac{24}{9}$, and $\dfrac{12}{24} = \dfrac{1}{2}$.

   b. Multiplying inverse operators gives 1; the system name is the product of the machine names.

3.

Input	Output
12	8
18	12
9	6
6	4
1	2/3

4. If $\frac{5}{9}$ of the teachers are female, $\frac{4}{9}$ are male. If $\frac{3}{8}$ of the males are single, then $\frac{3}{8} \cdot \frac{4}{9} = \frac{1}{6}$ are single males. If $\frac{1}{3}$ of the single males are over 50, $\frac{2}{3}$ are under 50. So $\frac{2}{3} \cdot \frac{1}{6} = \frac{1}{9}$ are single males under 50.

5. a. $\frac{3}{6}(8) = 4$ b. $\frac{3}{2}(2) = 3$ c. $4\left(\frac{2}{3}\right)(6) = 16$

6.

7. The first machine is a 1-for-2, so its label should be $\frac{1}{2}$. The second label is $\frac{1}{5}$. Even though their outputs are the same, they are performing different functions.

8. a. $\frac{1}{2} \cdot \frac{5}{6} \cdot 2 = \frac{10}{12} = \frac{5}{6}$ b. $\frac{2}{3} \cdot \frac{1}{6} = \frac{2}{18} = \frac{1}{9}$

c. $\dfrac{1}{2} \cdot \dfrac{1}{5} = \dfrac{2}{20} = \dfrac{1}{10}$ d. $\dfrac{1}{4} \cdot \dfrac{1}{3} = \dfrac{1}{12}$

9. 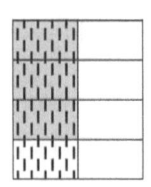 $\dfrac{1}{4} \cdot \dfrac{1}{3} \cdot \dfrac{1}{2} = \dfrac{1}{24}$

10. a. $\dfrac{3}{9} \cdot \dfrac{1}{3} \cdot \dfrac{1}{2} = \dfrac{1}{18}$ b. $\dfrac{2}{4} \cdot \dfrac{1}{2} \cdot \dfrac{1}{2} = \dfrac{2}{16}$ c. $\dfrac{1}{3} \cdot \dfrac{1}{2} = \dfrac{1}{6}$

11. a. $\dfrac{3}{8}$ 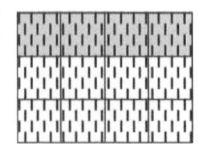   b. $\dfrac{4}{5}$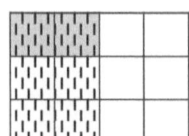

c. $\dfrac{1}{2}$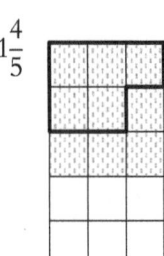

12. a. $1\dfrac{4}{5}$ b. $1\dfrac{1}{15}$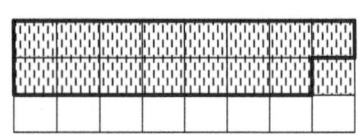

c. $4\dfrac{3}{8}$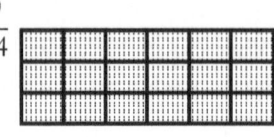

d. $1\dfrac{9}{24}$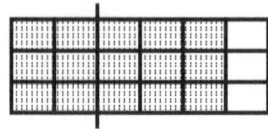

13. Ask yourself: what is the operator? Operator $= \dfrac{\text{output}}{\text{input}} = \dfrac{\text{ILS}}{\text{US}} = \dfrac{1}{0 \cdot 2239}$

Operator $\cdot$ 50 = 223.31 ILS

14. The operator is $\dfrac{0 \cdot 161}{1}$. The operator operates on 50 ZAR to give \$8.05 US.

15. The number of children is $\dfrac{1}{3}$ of the total, so the number of adults must be 32. The women are $\dfrac{1}{4}$ of the adults, so there are 8 women. That leaves 24 men.

16. What operator changes $\dfrac{360}{100}$ to $\dfrac{220}{100}$? $(\ ) \cdot \dfrac{360}{100} = \dfrac{220}{100}$

If we divide by 360 and multiply by 220, we get $\dfrac{220}{360} = \dfrac{11}{18} = 61.11\%$

17. a. 10 mm = 1 cm, so the operator is $\dfrac{1}{10}$ and $\dfrac{1}{10} \cdot 250 = 25\,\text{cm}$

b. 16 oz. = 1 lb, so the operator is $\dfrac{1}{16}$ and $\dfrac{1}{16} \cdot 224 = 14\,\text{lb}$

c. 1 minute = 60 sec, so the operator is $\dfrac{60}{1}$ and $65 \cdot \dfrac{60}{1} = 3900$ sec

d. 4 quarts = 1 gal, so the operator is $\dfrac{4}{1}$ and $22 \cdot \dfrac{4}{1} = 88\,\text{quarts}$

e. 100 cm = 1 m, so the operator is $\dfrac{1}{100}$ and $280 \cdot \dfrac{1}{100} = 2.8\,\text{m}$

18. No. It is not possible. For example, $4 \cdot 1\frac{1}{4} = 5$, but $6 \cdot 1\frac{1}{4} \neq 7$. Similarly for the other sizes. Both dimensions must be multiplied by the same factor in an enlargement.

19. Some possibilities are 4×5, $6 \times 7\frac{1}{2}$, and $2 \times 3\frac{1}{2}$.

20. Think in terms of what is paid. Yesterday you paid \$50. Today your sister will pay 90% of yesterday's price. So she will pay $\dfrac{9}{10} \cdot 50 = \$45$. Her savings totaled \$20.

21. a. Begin with 100. After a decrease of 10%, you have 90. 15% of 90 is 13.5. Therefore, you end up with 90 + 13.5 = 103.5 or 103.5% of your original amount.

b. 10% of 110 = 11. So you will have 110 − 11 = 99 or 99% of your original amount.

c. 60% of 50 = 30. You will have 50 + 30 = 80 or 80% of your original amount.

d. 50% of 120 = 60. You will have 120 − 60 = 60 or 60% of your original amount.

e.   25% of 70 = 17.5. You will have 70 + 17.5 = 87.5 or 87.5% of your original amount.

22.   After folding both papers, you should see $\frac{15}{18}$ and $\frac{8}{18}$, respectively.

23.   There are a number of possibilities. Here is one:

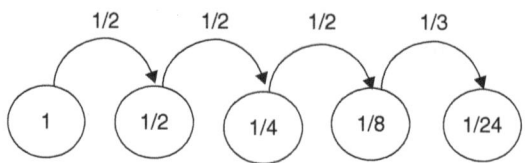

24.   a.   The operator is $\frac{6}{\frac{4}{3}} = \frac{18}{4}$. $\frac{18}{4}(6) = 27$ doughnuts.

b.   The operator is $\frac{\frac{2}{3}}{6} = \frac{1}{9}$. $\frac{1}{9}(21) = \frac{21}{9} = 2\frac{1}{3}$ pizzas.

c.   The operator is $\frac{10}{\frac{1}{4}} = 40$. $40(3) = 120$ peanuts.

d.   The operator is $\frac{2}{5}$. $\frac{2}{5}(6) = 2\frac{2}{5}$ peach pies.

25. $\frac{50}{100} \cdot$ _____ $= \frac{120}{100}$. Divide by 50 and multiply by 120 to get $\frac{120}{50} = \frac{240}{100}$ or 240%.

26.   Let's think in terms of how much you pay. After the sale price you pay 70% and when they take off another 25%, you pay 75% of yesterday's price. So $\frac{7}{10} \cdot \frac{3}{4} = \frac{21}{40} = \frac{42}{80} = \frac{52.5}{100}$ or 52.5%. This means you are saving 100% − 52.5% = 47.5%.

SUPPLEMENTARY ACTIVITIES

1.

Mark $\frac{1}{6}$ of the shaded part of this picture.

What is $\frac{3}{4} \cdot \frac{1}{6}$?

2.

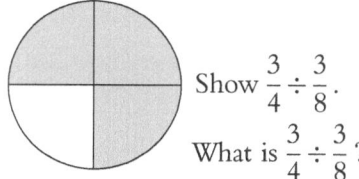

Show $\dfrac{3}{4} \div \dfrac{3}{8}$.

What is $\dfrac{3}{4} \div \dfrac{3}{8}$?

3. Reducing a unit to half its size and then tripling the result in size is equivalent to _____.

4. Dividing something into 5 equal shares and then quadrupling each share is equivalent to _____.

5. Make a picture showing a 2-for-3 machine with an input of 12 items and another picture of an 8-for-12 machine with an input of 12 items. What is the output of each machine? Name some other machines that will have the same output.

6. Draw a picture to illustrate the action of a 5-for-6 operator on a 3-for-4 operator on this set of objects:

7. Write the name of the shaded area as a product of fractions.

a. 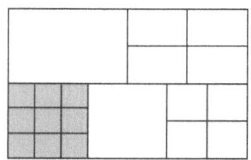   b.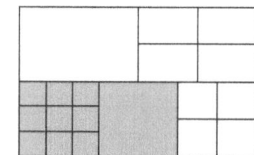

c.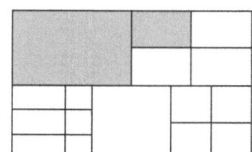

8. Find the number of children who are girls if you know that there are 12 male adults.

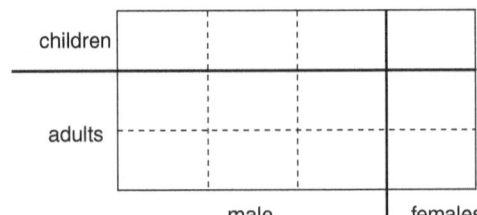

9. Fold one unit sheet of paper into thirds. Shade $\frac{1}{3}$. Fold another unit of the same size into fourths. Shade $\frac{1}{4}$. Open each unit and express the shaded amounts in twelfths. What conclusions can you draw?

10. Fold two paper units until you can rename each of these fractions with the same denominator:

$$\frac{3}{8} \text{ and} \frac{1}{6}$$

11. Complete each diagram and carry out the paper folding that will result in the smallest fraction.

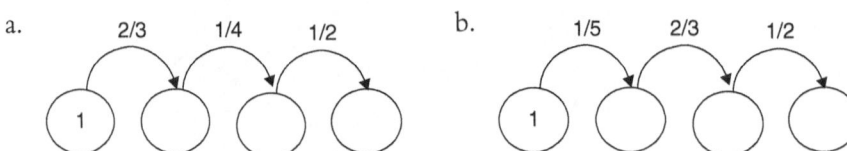

12. Can you see $\frac{3}{4}$ of $\frac{1}{2}$ of $\frac{1}{2}$? Shade it and then name the product $\frac{3}{4} \cdot \frac{1}{2} \cdot \frac{1}{2}$.

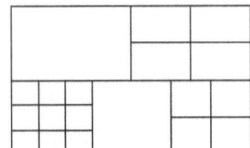

13. Can you see $\frac{2}{3}$ of $\frac{1}{3}$ of $\frac{1}{2}$? Shade it and then name the product $\frac{2}{3} \cdot \frac{1}{3} \cdot \frac{1}{2}$.

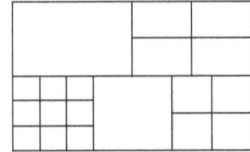

14. You have a photo that has been reduced to $\frac{3}{4}$ the size of the original. What percent do you need to use now to produce an image 125% larger than the original?

15. The department store had a sale yesterday and they advertised 30% off marked prices on all winter coats. Today they said they would take off an additional 25%. If you buy a coat today, what percent will you be saving?

16. What percent of the original quantity will result when

   a. A decrease of 20% is followed by an increase of 40%?
   b. An increase of 20% is followed by a decrease of 10%?
   c. A decrease of 20% is followed by an increase of 30%?
   d. An increase of 20% is followed by a decrease of 40%?
   e. A decrease of 30% is followed by an increase of 50%?

17. Write a complete multiplication statement ($a \cdot b = c$) based on each of these models.

   a. 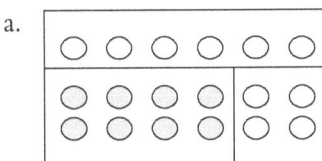   b.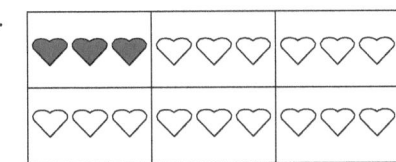

   c.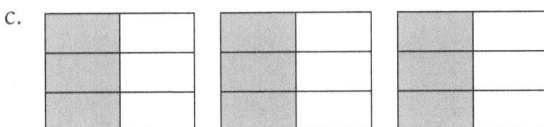

18. Use an area model for each problem.

   a. $\frac{5}{6} \cdot \frac{2}{5}$ b. $\frac{1}{5} \div \frac{1}{3}$ c. $1\frac{1}{3} \cdot \frac{3}{4}$ d. $\frac{1}{3} \div \frac{1}{6}$ e. $1\frac{1}{8} \div \frac{2}{3}$

   f. $2\frac{1}{2} \cdot \frac{4}{5}$ g. $1\frac{1}{4} \div \frac{5}{6}$ h. $2\frac{1}{2} \div \frac{1}{2}$ i. $1\frac{2}{3} \cdot \frac{4}{5}$ j. $\frac{3}{4} \div \frac{5}{8}$

19. A car manufacturer was discussing the possibility of installing several fuel-saving devices in a new car engine. One saved 50%, another saved 20% on fuel, and a third saved 30%. Of course, the expectation was that some might believe that the engine would not require any fuel at all! What is the real percent of fuel that could be saved is all three devices were installed?

20. For each problem, (a) find the operator and (b) solve the problem.

   a. Pat and Pam agree that a fair exchange is 3 apples for half a candy bar. If Pam gives Pat $3\frac{1}{2}$ candy bars, how many apples should he give her?

   b. Pat and Pam agree that a fair trade is $1\frac{1}{4}$ pounds of rice for $\frac{1}{2}$ pound of sugar. If Pat gives Pam 5 pounds of rice, how much sugar should he give her?

   c. Pat and Pam own livestock and they agree that a fair exchange is 2 cows for 9 sheep. If Pat has 26 sheep, how many cows can he expect to get from Pam?

21. A US dollar can be exchanged today for 108.275 Japanese Yen (JPY). If I exchange 5000 JPY for US dollars, how many dollars will I receive?

22. A South African rand (ZAR) is worth 0.1454 US dollars (USD). If I convert 300 USD, how many ZAR will I receive?

23. Fill in the missing information about this machine.

Input	Output
	12
9	
	4
1	

24. Someone forgot to fill in the label on this machine. Here is some information about what this machine is doing. Please label the machine.

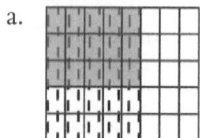

Input	Output
12 pieces	4 packages
18 pieces	6 packages
24 pieces	8 packages

25. Write the complete multiplication statement ($a \times b = c$) that can be read from each model.

a.    b.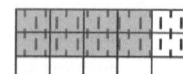

26. This figure shows a farmer's field which is a square, 1 mile on each side. There are 640 acres in a square mile. How many acres has the farmer used for each of the four crops shown in the drawing?

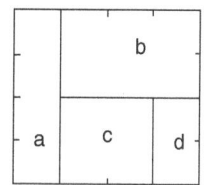

27. On your last visit to Germany, you paid 1.45 Euros per liter of gasoline. The American price at the time was $2.92 per gallon. (Note that 1 US gallon = 3.785 liters.) At these prices:

   a. How much would one gallon of European gas cost in dollars?
   b. How much would one liter of American gas cost in Euros?

28. You have a picture that measures $3\frac{1}{2}''$ × 5″. What is the largest that you can blow up the picture so that it still fits on a piece of standard paper that measures $8\frac{1}{2}''$ × 11″?

PRAXIS QUESTIONS

1. An appliance store gives a 15% discount off the list price on all merchandise, and an additional 30% reduction off the store price for a floor model. A TV set has a list price of $300 and the floor model sells for

   a. $210   b. $228.50   c. $178.50   d. 165   e. $135

2. What is the area of the shaded portion of this 6-meter square?

   a. 12 square meters   b. 22 square meters
   c. 20 square meters   d. 36 square meters
   e. 24 square meters

3. Jim got a discount of 20% off the marked price of a jacket that saved him $15. How much did he pay for the jacket?

   a. $35   b. $60   c. $75   d. $150   e. $300

4. Our shoe store had a sale, giving 25% off the regular price. Employees get an additional 15% of everything they buy. If Maria, a store employee, bought a pair of shoes marked $40, what did she pay for the shoes?

    a. $24   b. $27   c. $16   d. $25.50   e. $27.50

5. 1 Chilean Pesos = 0.00208 US Dollars. If I convert $20 US to pesos, how many pesos will I receive?

    a. 0.0416   b. 1.04   c. 9615   d. 4300   e. 4160

6. If you copied a document, making it 80% of its original size, at what percent must you copy that copy to return it to its original size?

    a. 100%   b. 20%   c. 120%   d. 125%   e. none of these

7. Multiplying 0.065 by which factor will *not* produce a larger product?

    a. $\dfrac{8}{7}$ b. $\dfrac{6}{5}$ c. 0.999   d. 1.003   e. $\dfrac{9}{8}$

8. You observe several people inserting quarters into a machine and receiving pieces of chocolate. Those results are given in the following table. If you insert $5.25, how many pieces of chocolate do you expect to receive?

# quarters	6	9	12
# pieces of chocolate	4	6	8

    a. 12   b. 14   c. 16   d. 18   e. 20

9. In a group of 36 people, $\dfrac{2}{9}$ of the men have blue eyes. $\dfrac{3}{4}$ of the people are men. How can you find the number of men with blue eyes?

    a. $\dfrac{1}{6}$ of 36   b. $\dfrac{2}{9}$ of 36   c. $\dfrac{5}{13}$ of 36   d. $\dfrac{35}{36}$ of 36

    e. not given

10. Which picture shows $\dfrac{1}{6}$ of 12?

    a. ●●●●●●
        ○○○○○○

    b. ●○○○○○
        ●○○○○○

    c. ●
        ●

    d. ○○●○○○
        ○○○○○○

    e. none of these

11. Jake went to work for $x per week. After several months, the company gave all of its employees a 10% pay cut. A few months later, the company gave everyone a 10% raise. What is Jake's new salary?

    a. $x   b. $0.09x   c. $0.99x   d. $1.01x   e. $1.11x

12. At 300% of its original size, a logo from my pizza box measures 4 inches square. How many inches square was the original?

    a. 2   b. $1\frac{1}{4}$ c. $2\frac{2}{3}$ d. $1\frac{1}{3}$ e. $1\frac{1}{2}$

13. If 1 GBP (Great British Pound) = 1.59 US Dollars, then 1 US dollar = how many GBPs?

    a. 0.6289   b. 0.41   c. 1.59   d. 1.69   e. not given

14. Two years ago you had to take a 22% pay decrease. This year, your employer has promised that your pay will be increased to its previous level. Approximately what should your increase be this year?

    a. 22%   b. 78%   c. 28%   d. 26%   e. 122%

15. 75% of Jake's salary of $5500/month is budgeted for tuition, books, rent, and food. $\frac{1}{3}$ of what remains is pocket money. How much pocket money does he have per month?

    a. $\frac{1}{3} * (0.25 * 5500)$ b. 10% of his income

    c. $\frac{1}{3} * (0.75 * 5500)$ d. $\frac{2}{3} * (5500 - 0.25 * 5500)$

    e. none of these

## CHAPTER 10

Fractions as Measures

DISCUSSION OF ACTIVITIES

1. a. $\frac{1}{2}$ of $\frac{1}{12}$ is $\frac{1}{24}$ and $\frac{1}{4}$ of $\frac{1}{24}$ is $\frac{1}{96}$.

    b. $\frac{1}{4}$ of $\frac{1}{12}$ is $\frac{1}{48}$ and $\frac{1}{2}$ of $\frac{1}{48}$ is $\frac{1}{96}$, and so on. Get $\frac{1}{192}$.

2. a. By successively partitioning, you can see that you have about $\frac{23}{32}$ of a tank. Answers may vary due to accuracy in partitioning, but you should get a result close to 0.71.

    b. $\frac{7}{16}$ tank or close to 0.44

3. a. You have $\frac{10}{16}$ or $\frac{5}{8}$ of a gallon left.

    b. You were able to buy about 5 gallons. Because the tank holds 14 gallons, that is $\frac{5}{14}$ tank.

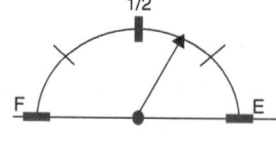

    c. By successively partitioning, you can see that you have about $\frac{3}{32}$ of a tank left. (Again, due to inaccuracy of partitioning, you may get a slightly different result, but it should be in the ball park of 0.09.) This means that you used $\frac{29}{32}$ of a tank, which was about 11 gallons of gas ($340 \div 31 = 10.967$). So if you divide the 11 gallons divided into 29 parts, then you would have $\frac{1}{32}$ of a tank, and 32 times that amount would be a full tank. $\frac{11}{29}(32) = 12.14$ gallons.

4. a. By successive partitioning, get $\frac{23}{32}$ or 27 ounces.

    b. By successive partitioning, get $\frac{9}{16}$ or 18 ounces

5.

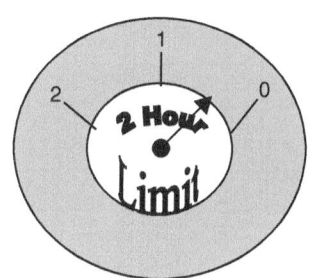

6. a. Partition into 24^{ths}. The order is $\dfrac{5}{6}, \dfrac{21}{24}, \dfrac{11}{12}$.

    b. Partition into 18^{ths}. The order is $\dfrac{3}{9}, \dfrac{1}{2}, \dfrac{5}{6}$.

    c. Partition into 28^{ths}. The order is $\dfrac{6}{7}, \dfrac{13}{14}, \dfrac{27}{28}$.

7. a. Divide your resulting fractions. As decimals, they should be between 0.67 and 0.71 (or close to these).

    b. Divide your resulting fractions. As decimals, they should be between 0.189 and 0.227 (or close to these).

8. a. $\dfrac{6}{9} = \dfrac{2}{3}$ b. $\dfrac{9}{20}$

9. a. One full rotation means that 12 hours have elapsed.

    b. 1 hour of time

    c. 12 minutes or $\dfrac{1}{5}$ hour

10. a. One full rotation means that 1 hour has elapsed.

    b. 5 minutes

    c. 1 minute

11. Annie says that by dividing 7 by any number between 11 and 12, her fractions will remain between $\dfrac{7}{12}$ and $\dfrac{7}{11}$. To keep your fractions between $\dfrac{7}{9}$ and $\dfrac{7}{8}$, divide 7 by any number between 8 and 9, such as $8\dfrac{1}{2}, 8\dfrac{1}{3}, 8\dfrac{1}{4}$. Check your fractions by dividing. They should be greater than 0.77 and less than 0.875.

12. Jon thought of 12 as $11\dfrac{4}{4}$, then by dividing 7 by any number between $11\dfrac{0}{4}$ and $11\dfrac{4}{4}$, he could write a fraction between $\dfrac{7}{12}$ and $\dfrac{7}{11}$. To keep your fractions between $\dfrac{7}{9}$ and $\dfrac{7}{8}$, divide 7 by any number between $8\dfrac{0}{4}$ and $8\dfrac{4}{4}$, such as $8\dfrac{1}{4}, 8\dfrac{1}{2}, 8\dfrac{3}{4}$. Check your fractions by dividing. They should be greater than 0.77 and less than 0.875.

13. a. By partitioning the 2 hours into 16^{ths}, you can see that $\frac{7}{8}$ of an hour is left, which is $52\frac{1}{2}$ minutes.

   b. By partitioning 1 hour into 8^{ths} or 2 hours into 16^{ths}, you can see that $\frac{3}{8}$ of an hour or $\frac{3}{16}$ of 2 hours remains. This is $22\frac{1}{2}$ minutes.

14. Don has partitioned the bottle's view strip to find that $4\left(\frac{1}{6}s\right)$ of the full bottle of oil remains. He knew that 1 whole bottle contained 32 ounces, so he substituted 32 for 1 whole. It doesn't appear that Don could evaluate $4\left(\frac{32}{6}\right)$, but he has certainly worked out the problem correctly. As a "next step," his teacher might urge him to work out how many ounces are equivalent to $4\left(\frac{32}{6}\right)$.

15. 2 spaces $= \frac{1}{4}$, so 8 spaces $= 1$. 1 spaces $= \frac{1}{8}$ and 5 spaces $= \frac{5}{8}$.

16. 4 spaces $= \frac{1}{3}$, so 12 spaces $= 1$. Cut every space in half to get 24^{ths} and then you can locate $\frac{7}{24}$.

17. a. Locate 1 and you will have eighths. 2 spaces $= \frac{1}{4} \cdot \frac{7}{8} > \frac{3}{4}$.

   b. $3\frac{1}{2}$ spaces $= 1$. Cut each space in half; then locate $\frac{4}{7}$. Partition again to locate $\frac{9}{14} \cdot \frac{4}{7} < \frac{9}{14}$.

   c. 2 spaces $= \frac{1}{6}$, so each space is $\frac{1}{12}$. After locating $\frac{7}{12}$, cut each space in half and locate $\frac{15}{24} \cdot \frac{7}{12} < \frac{15}{24}$.

   d. 6 spaces $= \frac{1}{5}$, so 30 spaces $= 1$ and 10 spaces $= \frac{1}{3}$. Partition each space into 3 equal lengths to get 15^{ths}. $\frac{1}{3} < \frac{7}{15}$.

18. a. At $3:18$, the minute hand will point to the third hash mark after the 3, and the hour hand will hit the half way mark in the second space after the 3.

   b. At $7:27$, the minute hand will point to the second hash mark past the 5, and the hours hand will be $\frac{1}{4}$ of the way between the third and fourth spaces after the 7.

c. At 10:52, the minute hand will be on the second hash mark after 10, and the hour hand will be $\frac{1}{3}$ of the way between the fourth and fifth spaces after the 10.

19. You can locate fractions with denominators of 2, 3, 6, 9, or 18.

SUPPLEMENTARY ACTIVITIES

1. Partition a number line and use arrow notation to find two fractions between those given.

    a. $\frac{1}{6}$ and $\frac{1}{5}$ b. $\frac{9}{10}$ and 1

2. On the clock at the right, draw the location of the hour hand and the minute hand when the time is precisely 1:25. Explain how you knew where to point the hands.

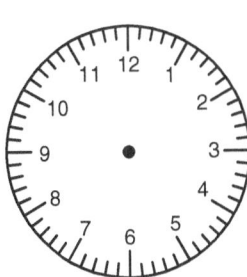

3. 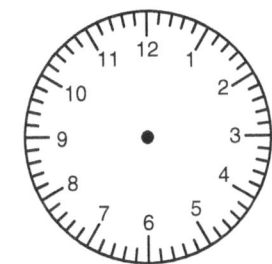 On the clock at the left, draw the locations of the hour and the minute hands when the time is 8:50.

4. The first gas meter shows how much gas you had before your trip and the second one shows how much you had when you got home. What part of a tank did you use for the trip? If your gas tank holds 16 gallons, how much gas did you use?

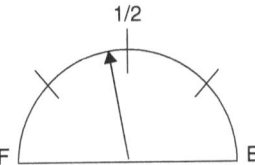
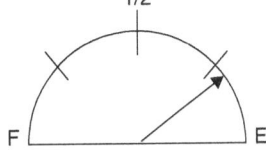

5. How much oil is left in each bottle?

a.

b.

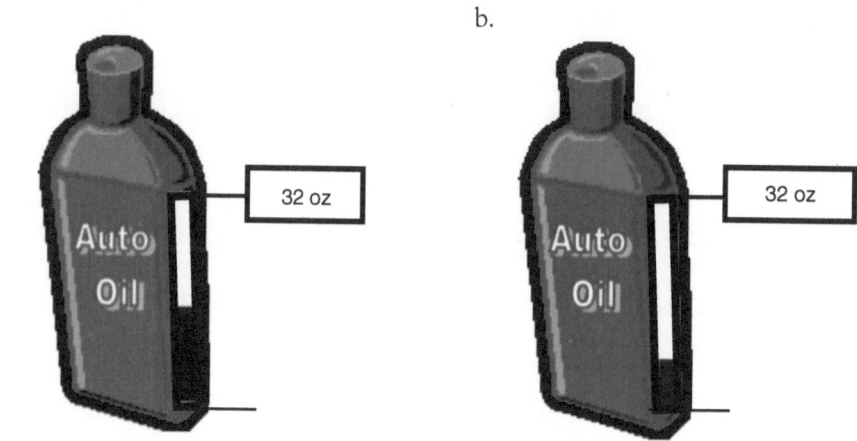

6. My book has 1004 pages and I am almost finished reading it. About how many pages have I read?

7. I have two stacks of $1000 bills. If the tall stack is worth $1 million, how much money is in the shorter stack? (Measure where the wrapper goes around the money.)

8. Tell how much time is left on each parking meter.

a.

b.

9. Parking downtown is expensive. The meters indicate that 1 nickel buys 3 minutes of time, for 1 dime you get 8 minutes, and for 1 quarter you get 20 minutes. You put in 2 nickels and 4 dimes. Draw an arrow to show how much time registers on the meter.

PRAXIS QUESTIONS

1. Which of the following could be about 25 centimeters long?

    a. a human thumb   b. a notebook   c. a car
    d. a house   e. a doorway

2. Arrange the following numbers as they would appear from left to right on a number line.

$$A = \frac{1}{100} \quad B = \frac{1}{10000} \quad C = 1 \quad D = 0.001 \quad E = \frac{1}{10}$$

    a. CEADB   b. BDAEC   c. none of these
    d. BAEDC   e. CBDAE

3. The number of grams in one kilogram is

    a. 0.001   b. 0.01   c. 0.1   d. 10   e. 1000

4. Which of the following measurements is not equal to the others?

    a. 230,000 millimeters   b. 0.23 kilometers   c. 23 meters

    d. 23,000 centimeters   e. 2.3 hectometers

5. The hour hand on a clock moves from one small hash mark to another. How many minutes have elapsed?

    a. 60   b. 5   c. 20

    d. 12   e. 15

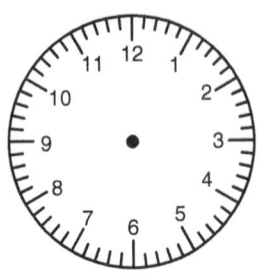

6. This thermometer measures degrees Celsius. What is the current temperature?

    a. 20.1   b. 20.2   c. 21

    d. 20.5   e. none of these

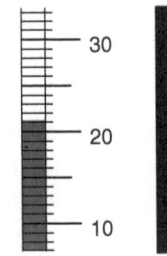

7. Which of the following measures could not be the height of a real building?

    a. 3.5 km   b. 60,000 mm   c. 23 cm   d. 29 m   e. 0.3 km

8.

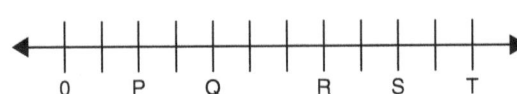

The number line above shows the relationships of the points P, Q, R, S, and T. The product of Q and R is 1.00. What do we know about the points on the number line?

    a. points Q and R are less than 1
    b. only point Q is less than 0
    c. point Q is more than 2
    d. the product of P and Q is less than 1
    e. the product of P and Q is between 1 and 2

9. Choose the point that would not lie between $\frac{5}{8}$ and $\frac{7}{8}$ on a number line.

    a. $\frac{37}{56}$ b. $\frac{19}{32}$ c. $\frac{63}{80}$ d. $\frac{37}{48}$ e. $\frac{11}{16}$

10. Which length is between $2\frac{1}{16}{}''$ and $2\frac{5}{8}{}''$ on a ruler?

    a. $2\frac{15}{32}{}''$ b. $2\frac{13}{16}{}''$ c. $2\frac{3}{4}{}''$ d. $2\frac{11}{16}{}''$ e. none of these

11. The cost at a parking meter is \$1.50 for 2 hours. If you put in \$0.60, how much time will register on the parking meter that shows $\frac{1}{8}$-hour intervals?

    a. between $\frac{1}{8}$ and $\frac{1}{4}$ hour   b. between $\frac{7}{8}$ and 1 hour

    c. between $\frac{5}{8}$ and $\frac{3}{4}$ hour   d. between $\frac{1}{2}$ and $\frac{5}{8}$ hour

    e. between $\frac{1}{4}$ and $\frac{3}{8}$ hour

12. If a steel bar is 0.39 feet long, its length in inches is

    a. less than 4   b. between 4 and $4\frac{1}{2}$ c. between $4\frac{1}{2}$ and 5

    d. between 5 and 6   e. more than 6

13. The length of a road was measured in kilometers. The length measured in meters would be

    a. one hundredth as much
    b. one hundred times as much
    c. one-tenth as much
    d. one-one thousandth as much
    e. none of these

14. How many ml of water are in this buret?

    a. 20.38   b. 23.8   c. 24

    d. 21.6   e. 23

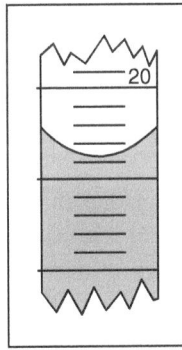

15. Which fraction could be located at point A?

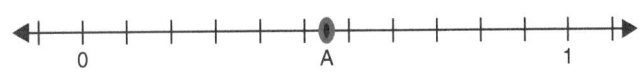

    a. $\frac{9}{20}$ b. $\frac{23}{48}$ c. $\frac{27}{48}$ d. $\frac{19}{44}$ e. $\frac{21}{44}$

## CHAPTER 11

Ratios and Rates

DISCUSSION OF ACTIVITIES

1. a. 2 boys : 3 girls

   b. The number of cows is $\frac{4}{5}$ the number of pigs, so the ratio of cows to pigs is 4 : 5.

   c. The ratio of Mary's height to her mom's height is 2 : 3.

   d. The ratio of Dan's weight to Becky's weight is 5 : 2.

2. a. $\frac{2}{7}$ b. 2 : 5

3. a. $\frac{80}{100}$ b. 30 balcony : 70 floor

   c. 20 empty : 80 occupied   d. 20 empty : 10 occupied in the balcony

4. a. The small gear makes $1\frac{3}{8}$ turns every time the large gear turns around once.

   b. When the small gear has made 5 turns, the large gear has made $3\frac{7}{11}$ turns.

   c. 4 turns of the small gears uses 32 teeth, so the large gear would need $32 \div \frac{4}{3}$ or 24 teeth.

   d. 4 turns of the large gear would require 44 teeth, so the small gear would need $44 \div \frac{11}{3} = 12$ teeth.

5. A, C, and F are correct.

6. a. 3 : 4 is greater than 5 : 9 or you have a better chance of winning with the odds 3 : 4 than with the odds 5 : 9.

   b. The picture shows that 3 : 4 = 15 : 20 and 5 : 9 = 15 : 27. This means that both ratios show 15 for, but 5 : 9 has 7 more against.

   c. $3(5 : 9) - 5(3 : 4) = (0 : 7)$

7. Other pictures are possible, depending on which fraction or ratio you clone.

   a. 11 : 12 is greater because, after removing copies of 5 : 6, we have 1 : 0. $11 : 12 - 2(5 : 6) = 1 : 0$

b. $12:16$ and $3:4$ are equivalent because cloning one produces the other. $4(3:4) = 12:16$:

c. Cloning $5:8$ and removing 5 copies of $7:9$, we see that we get 0 for:11 against, so $5:8$ is less than $7:9$. $7(5:8) - 5(7:9) = 0:11$

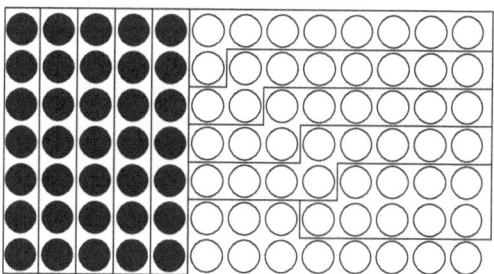

d. Make 5 copies of $\frac{7}{8}$ and 4 copies of $\frac{9}{10}$. The clone of $\frac{7}{8}$ is $\frac{35}{40}$. Rearranging the clone of $\frac{9}{10}$, we can see that it is $\frac{36}{40}$. Therefore $\frac{9}{10} > \frac{7}{8}$.

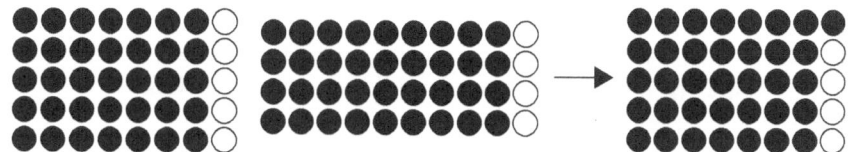

e. Make 9 copies of $7:8$ and remove 7 copies of $9:10$ to get $0:2$.

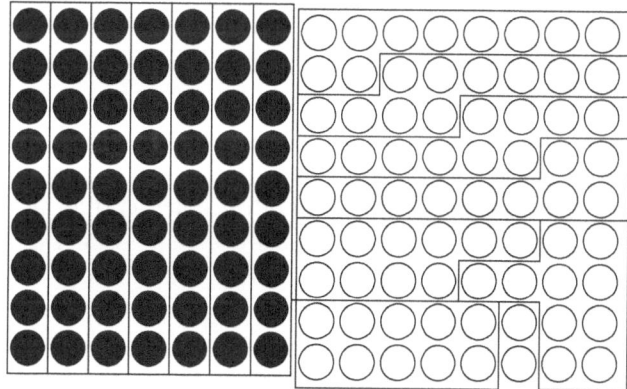

f. Clone $\frac{4}{9}$ and $\frac{3}{10}$ until there are 90 in each set. Rearrange the $\frac{4}{9}$ clone to get $\frac{40}{90}$ and compare to the $\frac{3}{10}$ clone, which is $\frac{27}{90}$. $\frac{4}{9} > \frac{3}{10}$.

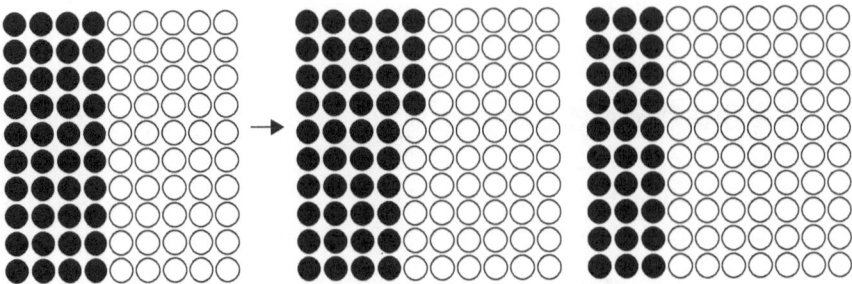

g. Removing copies of $4:5$ from $8:15$, we get $0:5$. Therefore, $4:5$ is larger. $8:15 - 2(4:5) = 0:5$.

8. a. The ratio $3:6$ is extendible over the people who are in the theater. Because the information about how many adults and how many children is already lost, we can further reduce this ratio, say, to $1:2$.

b. The ratio was not extendible, and not reducible without loss of information. We cannot add to or subtract teeth from a gear. The ratio of teeth on the small gear to the number of teeth on the large gear is $30:45$. Writing the ratio as $2:3$ gives up size of the gear.

c. The ratio $12:14$ is not extendible. We cannot add children to the classroom or increase the number of existing pets. In reducing the ratio, we lose the information about how many children are in the class.

d. The ratio is not extendible. For example, if the perimeter is 24, the area is 36, not 18. The ratio is also not reducible without losing information.

e. The ratio is extendible within the same bag of candies and to the extent that we don't go over the total number of pieces in the bag. The ratio is also reducible. For example, $3:6 = 1:2$. Information about the number of candies has already been lost.

f. The ratio is not extendible. It will not be true again that the ratio of the child's age to the mother's age will be $4:6$. The ratio is reducible to $2:3$ without loss of information, because $4:6$ does not give us present ages.

g. The ratio is not extendible. Dave will not be growing! If the ratio is reduced, we lose the information about both people's current heights.

h. The ratio $\$2.6:4$ oz. is extendible. We may assume that you can purchase more than 4 ounces at the same rate. The ratio is reducible to $\$0.65$ per ounce.

i. The ratio is not extendible. The family has only 6 children. If the ratio is reduced, we lose the information about how many children are in the family.

9. The ratios are realistically only approximations, but if we assume that they are exact and that the number of students is not changing, then the number of students has to be divisible by both 30 and 25 because we cannot hire fractions of

teachers. $30 = 2 \cdot 3 \cdot 5$ and $25 = 5 \cdot 5$. So look at multiples of 150. If the school has 300 students, then it has 10 teachers, and $300 : 12 = 25 : 1$, so they would need to hire 2 more teachers. If the school has 450 students and 15 teachers, the 450:18 gives the required ratio and they would need to hire 3 more teachers. Similarly, if the school has 600 students and 20 teachers, $600 : 24 = 25 : 1$, so they would need 4 more teachers. The number of teachers needed is approximately 1 for every 150 students.

10. Student A used a correct strategy. Notice that after removing one copy of $3 : 4$ from $5 : 8$, you are left with $2 : 4$, which is less than $3 : 4$. Student B used a correct strategy, but did not know how to interpret the result. Because there are 2 "against" dots remaining, $3 : 4 < 5 : 6$. We get 0:2 and conclude that $5 : 6 > 2 : 3$ although there were 3 copies of $5 : 6$ and 5 copies of $3 : 4$.

11. Charlie actually got the same deal as they offered at the other store.

12. a. Looking at the ratios of money to people: $3(59 : 7) - 7(26 : 3) = (177 : 21) - (182 : 21) = (-5 : 0)$. This means that for the Golden Theatre, the money to people ratio is less, so the people to money ratio is more. Golden has the better price.
    b. Looking at win to loss ratios: $5(7 : 9) - 7(5 : 7) = (35 : 45) - (35 : 49) = (0 : -4)$. This means that for the $7 : 9$ record, the loss to win ratio is more, so the win to loss ratio is better. $7 : 9$ is the better record.
    c. Looking at the seat to people ratios: $14(6 : 5) = 6(14 : 11) = (84 : 70) - (84 : 66) = (0 : 4)$. This means that for the car, the people to seats ratio is higher, so the seats to people ratio is lower and the car is more crowded.

13. There are 3 times as many boys, so there must be 3 times as many girls. There are 24 girls.

14. A should pay 2 parts, B should pay 4 parts, and C should pay 6 parts of the bill. Since the bill is $60 per week, $60 \div 12$ gives the cost of one part. This means A should pay $10, B should pay $20, and C should pay $30 each week.

15. $1.89 per half liter = $3.78 per liter.

16. 0.75 meters per 1 step = 7.5 meters per 10 steps. It would clearly take more than 10 steps to cover a distance of 10 meters.

17. A has 54 teeth and B has 36 teeth. This means that in one rotation of gear A, B turns $1\frac{1}{2}$ times. So if A makes 5.5 turns, B makes 8.25 turns.

18. a. You are going 4 mph.
    b. You traveled for $\frac{1}{4}$ hr.
    c. You are going 21 mph.
    d. It took you 12 minutes.
    e. You are going 48 mph.

19.   a. Find the total distance around the track (mi or km) covered by each driver by counting how many times he went around the divide by the time (3 hours) to get each driver's speed (mph or kph).
      b. Time the drivers to see how long it took them to drive the 50 km and then divide that distance (km) by time (hours) to get km per hour.

20.   You must be doing 72 miles per hour. It took you 50 sec to go 1 mile. That means you did 0.1 mi in 5 sec or 1.2 mi in 60 seconds (1 minute). Your hourly speed would then be 72 mph.

21.   If it takes me 20 minutes $\left(\text{or } \frac{1}{3} \text{ hr} \right)$ to get there at 40 mph, then the distance must be $\frac{40}{3}$ miles. $\frac{40}{3}$ at 50 mph will take about 16 minutes.

22.   Both jets will get there at the same time. They both travel 1200 mph and will be there in 5.5 minutes.

23. $\frac{1}{5}$ mi in 10 seconds $= \frac{1}{50}$ mi per second $= \frac{60}{50}$ mi per min. $= \frac{3600}{50}$ mi per hour $=$ 72 mph.

24.   Jim can cut $\frac{1}{4}$ lawn in 1 hour. His brother can cut $\frac{1}{3}$ lawn in 1 hour. Working together, in 1 hour the boys can cut $\frac{1}{4}$ lawn $+ \frac{1}{3}$ lawn $= \frac{7}{12}$ lawn. This means that they can cut $\frac{1}{12}$ lawn in $\frac{1}{7}$ hour, and the whole lawn in $\frac{12}{7}$ hours$= 1\frac{5}{7}$ hours $=$ 1 hour and 43 minutes.

25.   There are 130 km between stations A and C. It took the train 2.25 times as long to go between B and C, so the distance must be 2.25 times as great as between A and B. The distance between B and C is 90 km.

26.   At the rate of 13 parts per hour, worker 1 makes 117 parts in 9 hours. That means worker 2 made 126 parts in 9 hours or 14 parts per hour.

27.   If your salary rose $2.50 per hour then stayed the same for 2 years, the total increase was $2.50 over 3 years. That is an average increase of $0.83 per hour for the 3-years period.

28.   a. The change is 24 miles in 29 minutes or $\frac{29}{60}$ hour. $24 \div \frac{29}{60} = 49.66$ mph.
      b. The change is 39 miles in 32 minutes, which is 73.13 mph.
      c. The change is 63 miles in 61 minutes which is 61.97 mph, or about 62 mph.

29. a. After 10 minutes, B was ahead of A and stayed in the lead throughout the hour.
     b. In 60 minutes, B covered 6 miles. B's rate was 6 mph. A covered 2 miles in 6 minutes. A's rate was 2 mph.
     c. When you divide total distance covered by the total time it took, you get average speed. So 6 mph and 2 mph are average speeds. Typically, knowing an average speed does not tell us what rate a person was moving in any particular interval. They could be moving a little faster in one interval and a little slower in the next. However, in this case, because A's and B's positions in the sketch are in a straight line, we know that each kept the same pace throughout the 60 minutes.

30. a. 40 km in 3 hours = $13\frac{1}{3}$ kph
     b. 20 km in 2 hours = 10 kph
     c. 60 km in 5 hours = 12 kph

31. Walking takes three times as long as traveling by bus. In 8 hours, there are 4 times periods, each 2 hours long, so we need 2 hours of time on the bus, and 6 hours (3 times as long) to walk back. Riding for 2 hours at the rate of 9 mph on the bus, we can go 18 miles.

32. Troy has to run 40 m at a speed of 3 meters per second. He can do it in 20 seconds. Tara has to run 100 m and at 5 meters per second, she can do it in 20 seconds. The race should be a tie.

33. a. Jenny drove 3 times as far as they had already gone, and traveled at the same speed, so it should take her another 75 minutes to get home.
     b. Time and distance are proportional. The speed is the constant of proportionality. So k = 50 kilometers per 25 minutes = 2 km per min. So $t = d \div k = \frac{150}{2} = 75$ minutes.

34. a. Remember that the slope of a distance–time graph is speed. Both cars started at the same time, but one of them was traveling faster, so it covered a greater distance.
     b. One car started ahead of the other and traveled more slowly, so eventually the faster car met the slower car (the intersection point on the graph) and passed it. The faster car traveled a greater distance.
     c. One car started ahead of the other and because they both traveled at the same speed, stayed the same distance ahead of the other car as it was at the start.

35. Look at each one-hour interval on the graph to locate the steepest slope. During the second hour of the flight, the average speed of the plane was the greatest.

36.   Make a chart showing each runner's distance after 1 sec, 2 sec, etc. You will see that they must meet sometime between the third and fourth seconds of the race. They will meet precisely at 3.5 seconds into the race at a distance of 24.5 meters.

SUPPLEMENTARY PROBLEMS

1.   Sunnybrook Academy reported that their student–teacher ratio was 11 : 1. The public school in the same area has 624 students.

   a.   Why does Sunnybrook advertise the ratio 11 : 1 instead of reporting actual numbers of students and teachers?
   b.   How many teachers would the public school need to keep up with the academy?

2.   In a particular metal alloy, there are 1 part mercury to 5 parts copper, 3 parts tin to 10 parts copper, and 8 parts zinc to 15 parts copper.

   a.   What is the ratio of mercury to tin?
   b.   What is the ratio of zinc to tin?

3.   Steve compared 3 : 7 and 6 : 11 and his dot picture is shown.

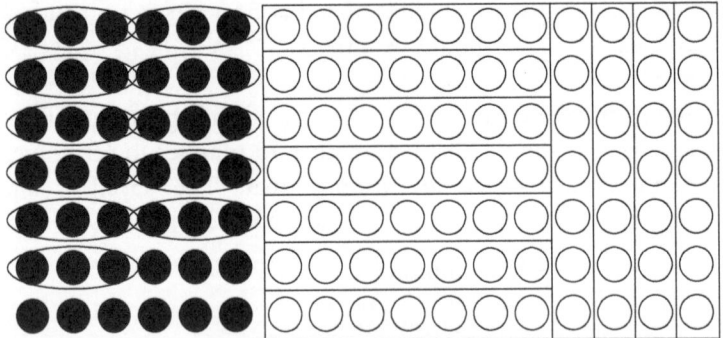

   a.   What do the 9 uncircled black dots mean?
   b.   Reinterpret the picture in terms of a ratio subtraction expressed symbolically.

4.   a.   Show two different dot pictures comparing 3 : 4 and 5 : 9.

yellow ----->   <--- red

b. For each of the pictures, express the difference between the ratios symbolically.

5. Using dot pictures, compare $5:7$ and $8:11$ two different ways, interpret each solution, and write the ratio subtraction symbolically.

6. Would you rather be taxed $15 on a purchase of $68 of $12 on a purchase of $52?

7. The Lamplight Theater sold 50 tickets in two weeks, while the Globe sold 150 tickets in 4 weeks. At which theater are the tickets selling faster?

8. The Jones Company has a loss of $500 in 3 weeks, while the Brown Company had a loss of $700 in 8 weeks. Which company suffered the least?

9. Which solution is stronger, 35 parts ammonia in 55 parts water, or 25 parts ammonia in 45 parts water?

10. How does one full turn of the yellow gear compare with one full turn of the red gear?

11. Here is Jennifer's work in which she compares $3:4$ and $5:8$. Is she correct?

Dear Mrs.L,

I can do it two ways.

If I start with $\frac{3}{4}$ I make clones until I can get $\frac{5}{8}$ out of it.

$\frac{3}{4}$ is bigger. There is money left and all the people got in.

If I start with $\frac{5}{8}$ I make clones until I can get $\frac{3}{4}$ out of it.

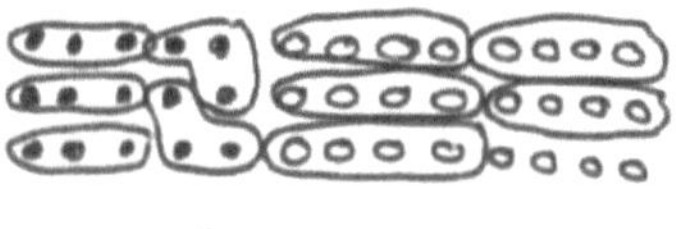

$\frac{5}{8}$ is smaller. There is no money left and 4 people still need to get in.

I go until I use up one of the numbers.

Jennifer

12. Suppose an object moves at a constant speed of 30 mph for ½ hour.

   a. Graph the relationship between its speed and time.
   b. How far did the object go during this time period?
   c. Where can this distance be seen on your graph?

13. Suppose a body moves 50 feet in 4 seconds in a straight line.

   a. If the body moved at a constant speed, what was its average speed for those 4 seconds?
   b. Suppose this object moved 10 feet in the first second, 10 feet in the second second, 20 feet in the third second, and 10 feet in the fourth second. What was the average speed during sec_1, sec_2, sec_3, and sec_4?
   c. Referring to part b: graph speed as a function of time from 0–4 seconds.

14. The activity of each of four people in their cars over a 6-hour period is recorded on these graphs. Tell what each person did.

a.
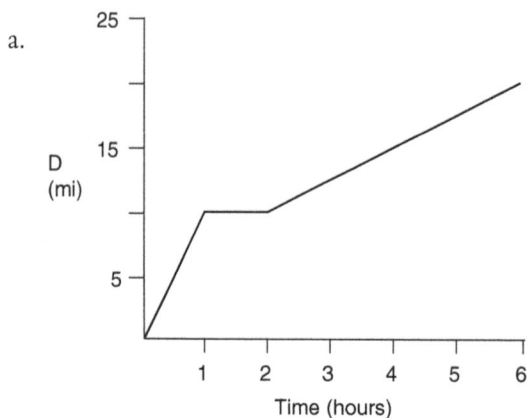

b.
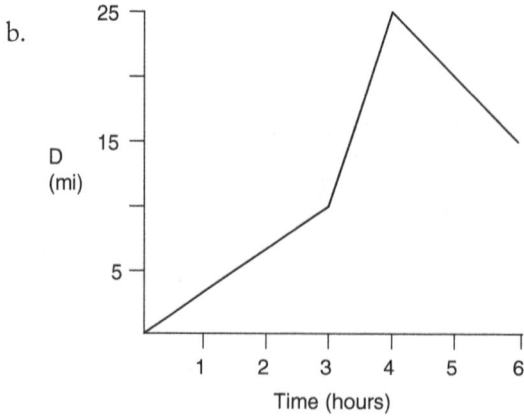

c.

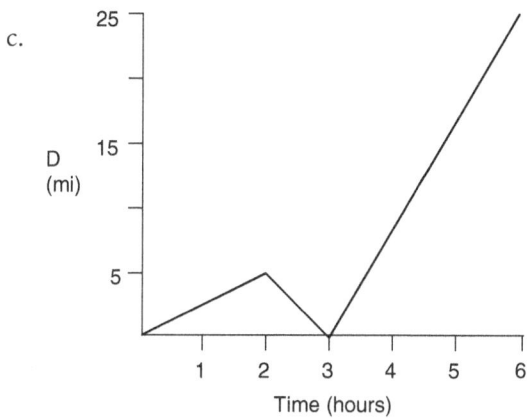

d.

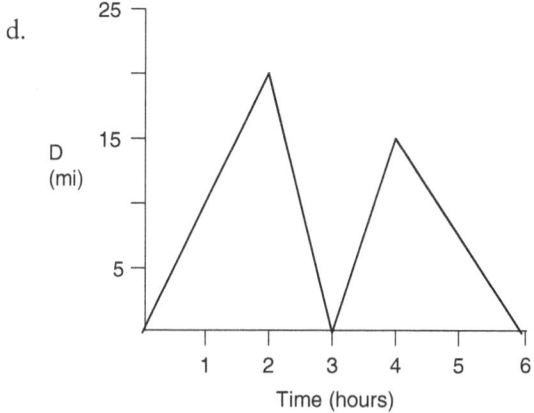

15. The following graphs show people's activity over a time period of 30 seconds. When possible, describe what each person is doing. If a graph does not make sense, tell why.

a.

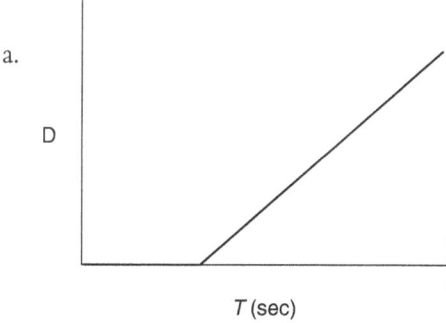

b.

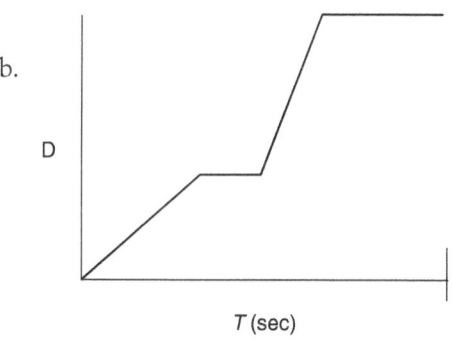

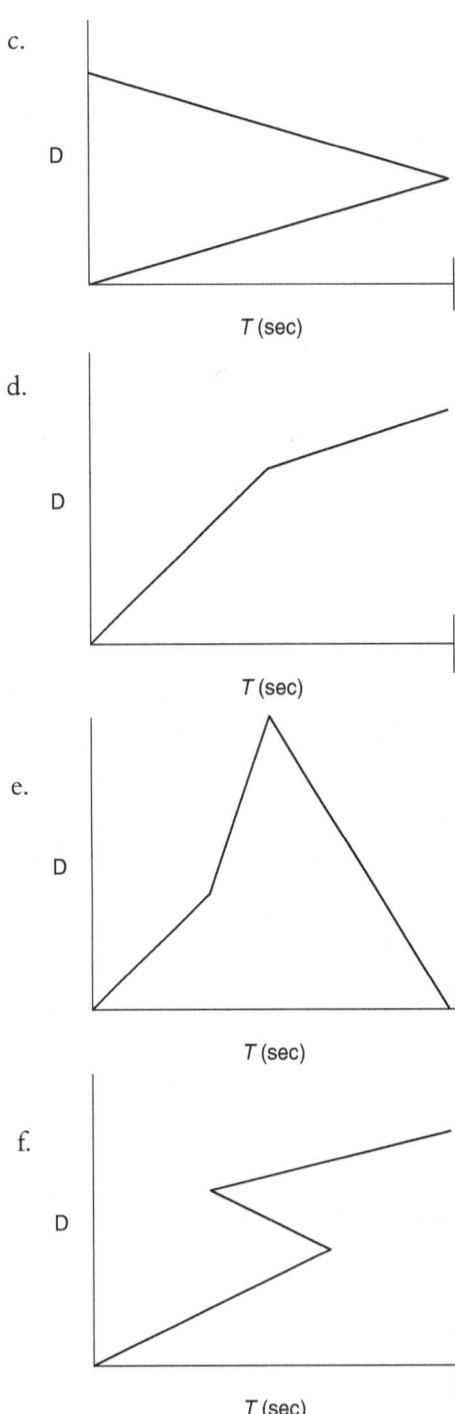

16. Sketch graphs showing the following information.

a. A car drives at a speed of 40 mph for 3 hours and then in the same direction at a speed of 60 mph for 2 hours.

b. A car drives for 1 hour at a speed of 50 mph then turns around and drives home at the speed of 40 mph.

17. Use the following graph of a car trip to answer these questions.

    a. How fast was this car moving during the first segment of its trip?
    b. During the second segment?
    c. During the third segment?
    d. What was its average speed for this 6-hour trip?
    e. Graph the average speed on the graph given above.

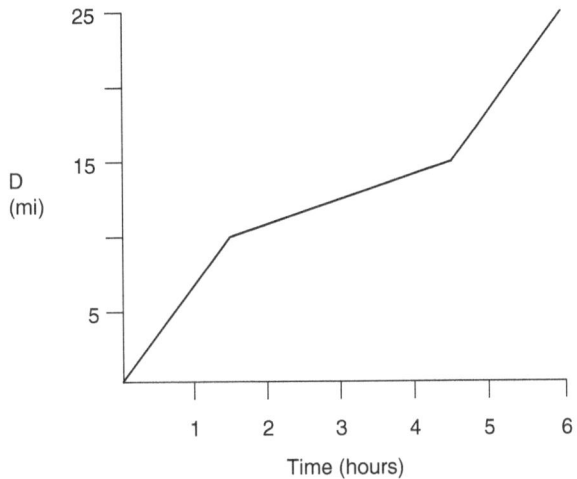

18. Sketch a graph that shows your activity.

    a. You run as fast as you can for 5 seconds, rest for 5 seconds, then walk for 20 seconds.
    b. You walk for 10 seconds, run as fast as you can for 10 seconds, and then turn around and walk back towards your starting point for 10 seconds.
    c. You walk for 10 seconds, turn around and walk back to where you started in 5 seconds, and then run as fast as you can away from your starting point for 15 seconds.
    d. You walk for 15 seconds, turn around and walk toward your starting point for 5 seconds, then turn and walk away from the starting point again and walk for 10 seconds.

19. Charlie got a traffic ticket for a moving violation: traveling 5 mph over the 55 mph speed limit. The officer's citation said that he followed Charlie for 10 miles and that he averaged 60 mph, which was 5 mph over the speed limit. Charlie drove at 80 mph for the first 5 miles and then at 40 mph for the next 5 miles. Charlie went to traffic court. How did he argue his case?

20. Frank drove from Boston to Washington and back again. He averaged 50 mph on the way and 60 mph on the way back. The round trip took him 18 hours. How far apart are the two cities?

21. A cruise ship headed out for a trans-Atlantic trip traveling at a constant speed of 25 knots. A passenger who missed the departure hired a speedboat to catch up. The speedboat traveled at 45 knots. The speedboat left the dock precisely 2 hours after the cruise ship did. How far from the dock were the two boats when they met?

22. Frank and Flo have a pet fly named Frieda, who flies back and forth between them at a rate of 50 mph while they are riding their bikes. Frank and Flo were 10 miles apart, biking towards each other, going 20 mph. Frieda Fly started on Frank's handlebars and flew along the path to Flo's handlebars, and then back to Frank's, then back to Flo's, until Frank and Flo met each other on the path. How far did Frieda travel?

23. A boy can bike a mile in 5 minutes and walk a mile in 20 minutes. How much time does he save if he bikes to his dad's office, 8 miles away, rather than walking?

24. Carrie climbed Crow's Peak at the rate of $1\frac{1}{2}$ miles per hour and came down at the rate of $4\frac{1}{2}$ miles per hour. It took her 6 hours to travel both ways.

     a. How far is the top of the peak?
     b. How long did it take her to get down?

25. Jim and Ken are brothers and they are both on the track team. They frequently race each other in practice, giving Jim a bit of an advantage because he is younger and his best speed is only 3 meters per second. Ken runs 5 meters per second. This week, they ran a 150-meter race, and Ken gave Jim a 30-meter head start. Which brother won the race? How far into the race did he pass the loser?

26. A bus travels 2 miles uphill and its average speed on the way up is 25 mph. At what speed would it have to travel down the hill so that the average speed for the entire trip is 40 mph?

27. Compare these rates by finding the constant associated with their equivalence classes.

     a. 30 heartbeats per 15 seconds; 64 heartbeats in 30 seconds.
     b. a loss of 30 pounds in 7 months; a loss of 8 pounds in 6 weeks.

PRAXIS QUESTIONS

1. If a train covers 14 miles in 10 minutes, then the rate of the train in miles per hour is

    a. 140   b. 112   c. 84   d. 100   e. 98

2. Two ships leave from the same port at 11:30 AM. If one sails due east at 20 mph and the other due south at 15 mph, how many miles apart are the ships at 2:30 PM?

   a. 25   b. 50   c. 75   d. 80   e. 35

3. Riding a bike to school takes 25 minutes. Driving home takes only 5 minutes. If the trip to school is 3 miles long, what is the average speed for the whole trip?

   a. 8 mph   b. 12 mph   c. 9 mph   d. 14 mph   e. 6 mph

4. It took the oil truck 3 hours to fill the empty oil tank pumping at 100 gallons an hour. How many hours will it take another oil truck to fill the same size tank pumping 80 gallons an hour?

   a. $2\frac{2}{5}$ b. $3\frac{3}{5}$ c. $3\frac{3}{8}$ d. $3\frac{3}{4}$ e. none of these

5. The cost of 30 onions is d dollars. At this rate, how many onions can you buy for 80 cents?

   a. $\frac{24}{d}$ b. $\frac{240}{d}$ c. $\frac{3d}{8}$ d. $\frac{8d}{3}$ e. $\frac{4d}{16}$

6. Joshua can mow a lawn in minutes. What part of the job can he do in 15 minutes?

   a. $t-15$ b. $\frac{t}{15}$ c. $15t$ d. $15-t$ e. $\frac{15}{t}$

7. If t tons of snow fall in 1 second, how many tons fall in m minutes?

   a. $mt+60$ b. mt c. $\frac{60m}{t}$ d. $\frac{mt}{60}$ e. none of these

8. $0.44 per 88 grams is the same price as

   a. $0.02 per gram   b. $0.005 per gram   c. $44 per 880 grams
   d. $0.04 per 11 grams   e. none of these

9. A pill contains 0.2 grams of a medicine. How many pills can be made from 1 kilogram of the medicine?

   a. 20   b. 200   c. 500   d. 5,000   e. 2,000

10. A man travels a certain distance at 60 mph and returns over the same road at 40 mph. What is his average rate for the round trip in miles per hour?

    a. 42   b. 44   c. 46   d. 48   e. 50

11. At the rate of 28 lines per page, a book has 300 pages. If the book has to contain only 280 pages, how many lines should a page contain?

    a. 29   b. 32   c. 30   d. 31   e. none of these

12. It took Ashley 12 hours to travel by train from New York to North Carolina at an average of 55 miles per hour. On the return trip, she traveled by bus and averaged 45 miles per hour. How many hours did the return trip take?

   a. $13\frac{2}{3}$ b. 14   c. $14\frac{2}{3}$ d. 15   e. 16

13. 50 mph is approximately 80 kph (kilometers per hour). If you are traveling 100 kph, your speed in mph is closest to

   a. 60   b. 70   c. 80   d. 90   e. 100

14. Gerald is paid a base salary of $60 plus a 5% commission. In a week in which his sales amounted to $600, the ratio of his base salary to his commission was

   a. 2 to 1   b. 1 to 2   c. 2 to 3   d. 3 to 2   e. 3 to 1

15. A hiker walks a distance of 16 miles at the rate of 3.25 miles per hour. If he doubles his speed, what will happen?

   a. He will save almost 2.5 hours.
   b. He will go 32 miles in the same amount of time.
   c. He will walk 13 miles in 2 hours.
   d. He will double his time.
   e. All of the above

## CHAPTER 12

Challenging Problems

1. Roxie and her dad went on a hiking trip together on Morecrest Mountain. She carried the supplies $\frac{1}{3}$ of the way up and $\frac{3}{5}$ of the way down. If she carried the supply pack a distance of $17\frac{1}{2}$ miles, how far did they hike up the mountain?

2. In class yesterday, Professor Fracto noticed that $\frac{3}{4}$ of his students were female and there were 6 people absent. When he checked his records, he found that $\frac{2}{3}$ of those enrolled were female. Has there been a mistake?

3. A group of friends ate a bag of pistachios in one sitting, leaving just one nut in the bowl for me. They did not worry about dividing the nuts equally among themselves because there were too busy chatting and some like pistachios more than others do. Mandy took $\frac{2}{7}$ of them and Trish, $\frac{1}{12}$ of them. Alison and Kristen ate $\frac{1}{6}$ and $\frac{1}{3}$ respectively. Carol took 20, Kate took 12, and Pam, who was so busy talking that she forgot to eat, took 11 with her when she left. How many nuts were in the bowl before the girls got to them?

4. Solve this puzzle. It is time for bed! One fifth of three-eighths of what remains of the morning has already gone by. What time is it (to the nearest second)?

5. A man died and left to his elder son the task of splitting the five large bills, his lifetime savings. The son gave his brother twice two-thirds of his own share, and his mother $\frac{2}{8}$ of the younger brother's share. What fraction of the money did each receive?

6. Jim's wife had a baby yesterday morning. When the baby was born, the time until 12 noon was six times $\frac{2}{7}$ of the time since midnight. At what time was the baby born?

7. Oscar, Harvey, and Merv each have huge collections of pennies. One day they decided to weigh them. Oscar discovered that his collection outweighed Harvey's

by $\frac{1}{3}$ of Merv's. Harvey's was the weight of Merv's plus $\frac{1}{3}$ of Oscar's. Merv's collection weighed 10 pounds more than $\frac{1}{3}$ of Harvey's. How many pounds of pennies did each boy have?

8. Last Christmas, Mr. Moore gave each of his six daughters a box of a different weight, the lightest for the youngest, and so on until the heaviest went to the oldest daughter. Each box was one ounce heavier than the previous. If the sum of the weights was 6 pounds, what were the 6 weights?

9. Before he ended his life, the man who is buried here composed the inscription for his grave marker so that you could figure out the age at which he died. How old was he?

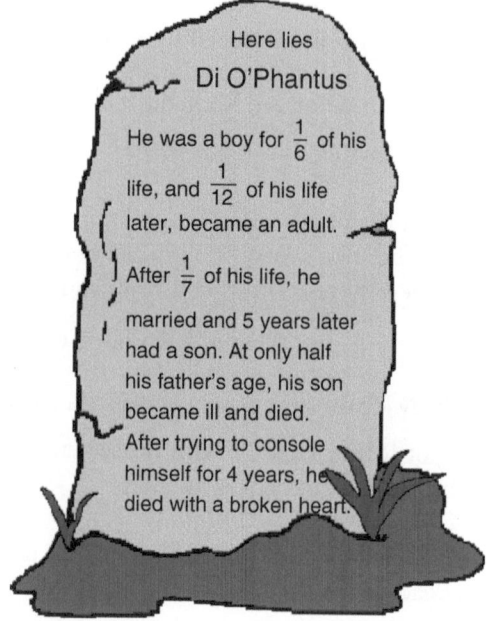

Here lies
Di O'Phantus

He was a boy for $\frac{1}{6}$ of his life, and $\frac{1}{12}$ of his life later, became an adult. After $\frac{1}{7}$ of his life, he married and 5 years later had a son. At only half his father's age, his son became ill and died. After trying to console himself for 4 years, he died with a broken heart.

10. A certain sculpture was made using 60 pounds of four different metals. The gold and bronze together accounted for $\frac{2}{3}$ of its weight. The gold and tin together comprised $\frac{3}{4}$ of its weight. The gold and iron together weighed 36 pounds. How much of each metal was used?

11. A man left this will. What was the dead man's fortune?
I leave my son $\frac{1}{5}$ of my wealth, and my wife, $\frac{1}{12}$. I leave to each of my four grandsons, my two brothers, and my grieving mother, $\frac{1}{11}$ of my property. To my cousins, I leave $12,000 to share and to my friend, Tom, $5,000. To my faithful

employees I leave the following bonuses for their service: Ann, $350; Mark, $350; Steve, $600; Ray, $300; the delivery man, $200; the janitor, $200. It is my wish that $25,000 be spent on my mausoleum. Use $9,000 to cover the other costs associated with a modest funeral.

12. When my brother and I added the mileage we did on our bikes, we got 20 miles. If you add a third of my distance and a fourth of my brother's, you get 6 miles, the distance our mother cycled. How far did I go? How far did my brother go?

13. On Thursday night at the Larson household, each person gets home at a different time. Mike was home first. He skipped dinner, opened a bag of candy bars, and ate 6 of them. Alison came home next and helped herself to $\frac{1}{3}$ of the original contents of the bag. Mark got home just before Jenny and had $\frac{1}{2}$ of the remaining candy bars for his dessert. Jenny finished off the last two candy bars. Later, when Mrs. Larson got home and asked who ate the candy bars, Alison, the youngest, got blamed for eating most of the bag. For being so inconsiderate, she had to use her own money to buy a new bag of candy bars. Was this fair?

14. The Belly Jean Company sells purple passion jelly beans for $1.25 a pound and pink pineapple jelly beans for $1.85 a pound. How many pounds of purple must be added to 50 pounds of the pink jelly beans to create a mixture that sells for $1.45 a pound?

15. A man was stranded on a desert island with enough water to last him 27 days. After 3 days, he saved a woman on a small life raft. If they can keep their water supply from evaporating, they figure that they can share their water equally for 18 days. What portion of the man's original daily ration was allotted to the woman?

16. In a certain scout troop, 70% of the scouts had a compass, 75% had matches, 85% had a knife, and 85% had a watch. What percent of the scouts had all four pieces of equipment?

17. At a recent foxhunt, Prince Charles timed his favorite hound and found that he was running 10 m for every 6 m run by the fox. At one point, a fox was 30 m ahead of the hound. How far did the hound have to run to catch up with the fox?

18. Did you ever walk the steps of a moving escalator? In my favorite department store, I found that if I walk down 26 steps, I can get to the bottom in 30 seconds, and if I walk down 34 steps, I can get to the bottom in 18 seconds. How many steps are in the moving stairway? (Time is measured from the moment the first step beings to descend until I step on to the solid platform at the bottom.)

19.   An army camp had enough food for 1200 people for 8 weeks. After 3 weeks, 300 more soldiers joined the camp. How long will the food last?

20.   There is a rectangular sheet of paper which, when cut in half, yields two smaller rectangles, each of which is similar to the original sheet. What is the ratio of the length to the width of the original piece of paper?

21.   Ram had 3 pieces of flat bread and Shyam had 5. A hungry traveler asked to share the bread and they cut it into 3 equal shares. After dinner, the traveler gave the men 8 coins for his share of the bread. How should Ram and Shyam divide the 8 coins?

22.   Sam, amateur detective, was walking along at the rate of about $3\frac{1}{2}$ mph when a car passed him on the road and tossed a gun out the window. He heard sirens back in town and immediately suspected that it was a get-away car. He counted his steps—29 in all—until the car turned the corner and he lost sight of it. It took him another 203 steps to get to the corner, but by then, the car was gone. He told the police that he was not sure it was the get-away car because it was not moving very quickly. How did Sam know how fast the car was moving?

Cuisenaire Strips

O = orange
Bl = blue
Br = brown
Bk = black
DkG = dark green
Y= yellow
M = magenta
LGr = lime green
R = red
W = white

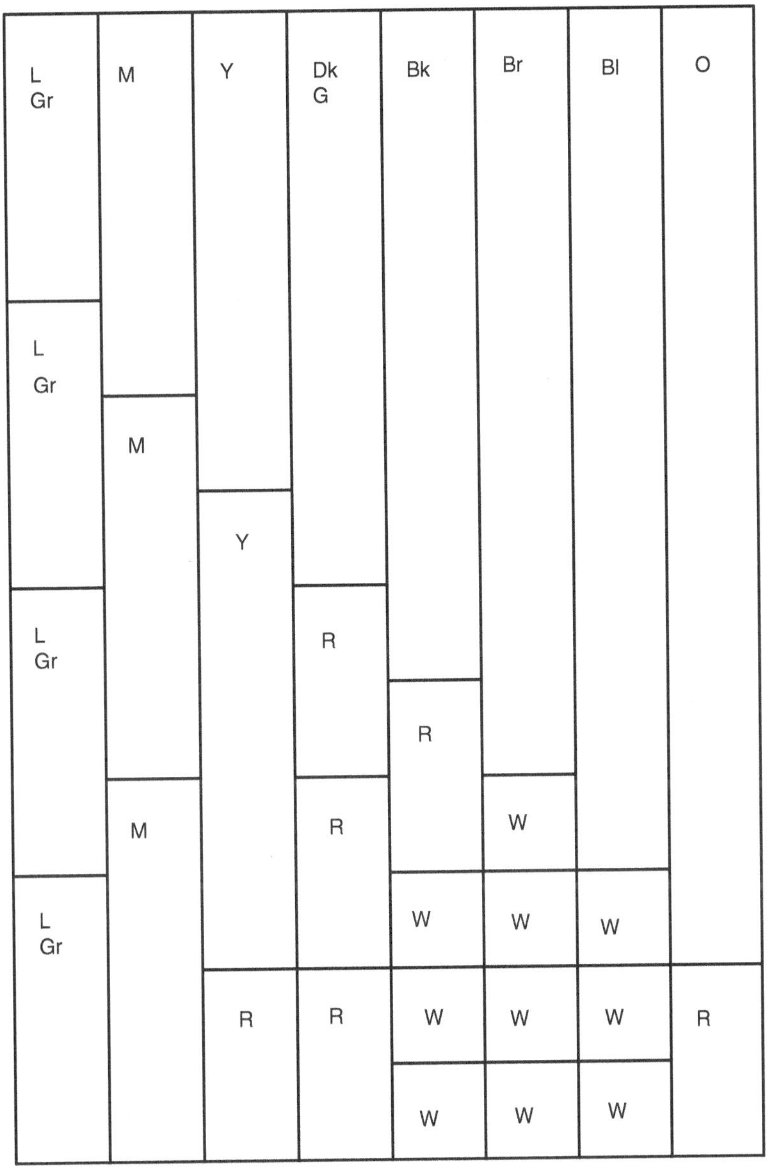

L Gr	M	Y	Dk G	Bk	Br	Bl	O
L Gr	M	Y	Dk G	Bk	Br	Bl	O
L Gr	M	Y	R	R	W		
L Gr	M	R	R	W	W	W	
L Gr	R	R	W	W	W	R	
				W	W	W	

Cut 2 copies out of card stock and color.

Pattern Pieces

Make multiple copies. You will need at least 2 hexagons, 4 trapezoids, 6 rhombuses and 12 triangles. Color your pieces according to the key below.

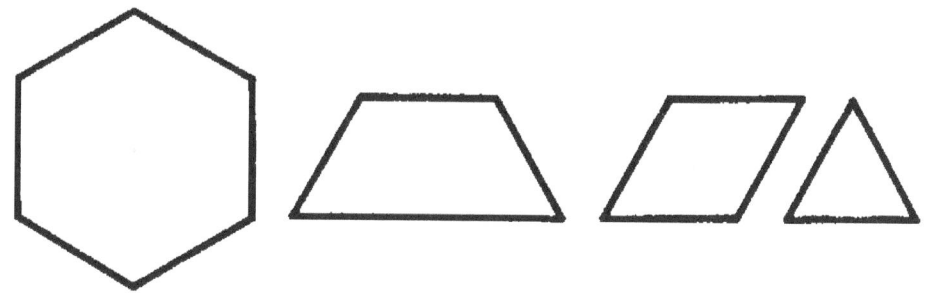

Hexagon—yellow
Trapezoid—red
Rhombus—blue
Triangle—green

Fraction Strips for Partitioning

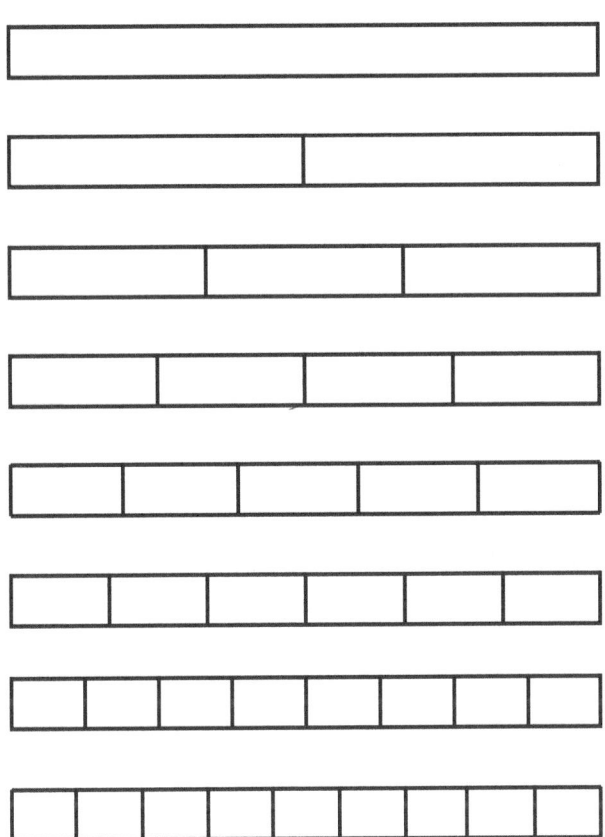